统计与大数据"十三五"规划教材立项项目
数据科学与统计系列规划教材

大数据
专业英语教程

An English Course of Big Data

张强华 司爱侠 朱丽丽 张千帆 ◎ 编著

人民邮电出版社
北京

图书在版编目（CIP）数据

大数据专业英语教程 ：附全套音频 / 张强华等编著
. -- 北京 ：人民邮电出版社，2021.1（2024.1重印）
数据科学与统计系列规划教材
ISBN 978-7-115-53854-3

Ⅰ. ①大… Ⅱ. ①张… Ⅲ. ①数据处理－英语－教材
Ⅳ. ①TP274

中国版本图书馆CIP数据核字(2020)第204153号

内 容 提 要

本书选材广泛，共 12 个单元，主要内容涉及什么是大数据，大数据分析，数据模型，结构化数据、半结构化数据和非结构化数据，大数据存储，数据提取、转换、加载，数据备份，Python 编程语言与 R 编程语言，数据库基本概念，数据库管理系统，数据仓库，云存储，数据处理，数据挖掘及其算法，Hadoop 与 Spark，大数据可视化，商业智能，数据安全，大数据和人工智能的成功关键等。

本书内容设置与课堂教学的各个环节紧密切合，支持备课、教学、复习及考试等教学环节。另外，本书为用书教师提供 PPT、参考答案、音频文件、教学大纲等资源。

本书既可作为高等院校本科、专科大数据领域相关的专业英语教材，也可供从业人员自学。

◆ 编　　著　张强华　司爱侠　朱丽丽　张千帆
　责任编辑　孙燕燕
　责任印制　周昇亮
◆ 人民邮电出版社出版发行　　北京市丰台区成寿寺路 11 号
　邮编　100164　　电子邮件　315@ptpress.com.cn
　网址　https://www.ptpress.com.cn
　固安县铭成印刷有限公司印刷
◆ 开本：800×1000　1/16
　印张：15　　　　　　　　　2021 年 1 月第 1 版
　字数：354 千字　　　　　　2024 年 1 月河北第 7 次印刷

定价：49.80 元

读者服务热线：(010)81055256　印装质量热线：(010)81055316
反盗版热线：(010)81055315
广告经营许可证：京东市监广登字 20170147 号

前言

党的二十大报告指出，要加快建设数字中国。习近平总书记深刻指出，加快数字中国建设，就是要适应我国发展新的历史方位，全面贯彻新发展理念，以信息化培育新动能，用新动能推动新发展，以新发展创造新辉煌。中共中央、国务院印发了《数字中国建设整体布局规划》，从党和国家事业发展全局和战略高度，提出了新时代数字中国建设的整体战略，明确了数字中国建设的指导思想、主要目标、重点任务和保障措施。建设数字中国是数字时代推进中国式现代化的重要引擎，是构筑国家竞争新优势的有力支撑。当今社会对大数据人才的需求可谓供不应求。为了培养符合社会需求的专业人才，我国许多高校纷纷开设与大数据相关的专业。由于大数据专业的国际化特征较为明显，社会对从业人员的专业英语水平提出了更高的要求。因此，具备相关职业技能并精通专业英语的从业人员更有机会占据竞争优势，成为职场中不可或缺的核心人才。为了贯彻落实二十大精神，我们编写了本教材。

本书的特点如下。

（1）选材广泛，内容严谨。本书对大数据基础知识、大数据常用操作、大数据编程与工具软件、大数据应用等内容进行了详细介绍，并且课文素材均经过严谨推敲与细致加工，以期更好地满足教学需求。

（2）体例新颖，适合教学。本书内容设置与课堂教学的各个环节紧密切合，支持备课、教学、复习及考试等教学环节。每一单元包含以下部分：课文——选材广泛、风格多样、切合实际；单词——给出课文中出现的新词，读者由此可以积累大数据专业的基本词汇；词组——给出课文中的常用词组；缩略语——给出课文中出现的、业内常用的缩略语；习题——包括针对课文的词汇练习、翻译练习和填空练习；参考译文——让读者对照理解并提高翻译能力。

（3）教学资源丰富，教学支持完善。本书有配套的 PPT、参考答案、音频文件（扫描二维码获取）、教学大纲、词汇总表等资源。另外，书中的习题数量适当，题型丰富，难易搭配，便于教师组织教学。

在使用本书过程中，如有任何问题，读者都可以通过 E-mail 与我们交流，我们一定会给予及时答复。如发送邮件索取本书配套教学资源，请在主题处注明姓名及“索取大数据专业英语参考资料”字样，我们的 E-mail 地址如下：zqh3882355@sina.com、zqh3882355@163.com。选用本书的教师也可到人邮教育社区（www.ryjiaoyu.com）免费下载课件。

编者

目　录

Unit 1

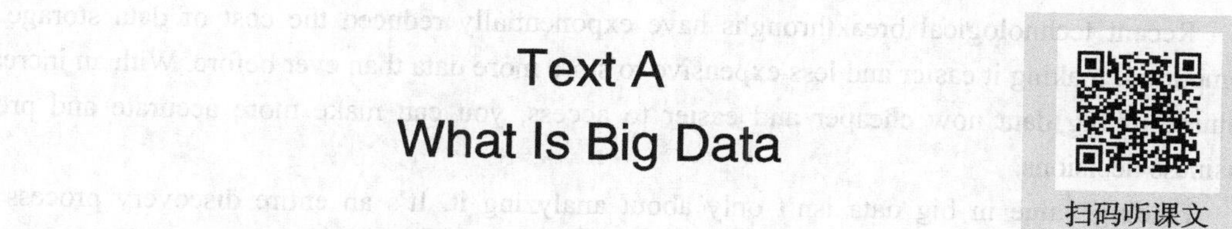

Text A
What Is Big Data

1. The Definition of Big Data

Big data is data that contains greater variety arriving in increasing volumes and with ever-higher velocity, according to Gartner. This is known as the three Vs.

To put it simply, big data is larger, more complex data sets, especially from new data sources. These data sets are so voluminous that traditional data processing software just can't manage them. But these massive volumes of data can be used to address business problems you wouldn't have been able to tackle before.

2. The Three Vs of Big Data

2.1 Volume

The amount of data matters. With big data, you'll have to process high volumes of low-density, unstructured data. This can be data of unknown value, such as Twitter data feeds, clickstreams on a webpage or a mobile app, or sensor-enabled equipment. For some organizations, this might be tens of terabytes of data, for others, it may be hundreds of petabytes.

2.2 Velocity

Velocity is the fast rate at which data is received and perhaps acted on. Some internet-enabled smart products operate in real time or near real time and will require real-time evaluation and action.

2.3 Variety

Variety refers to the many types of data that are available. Traditional data types were structured and fit neatly in a relational database. With the rise of big data, data comes in new unstructured data types. Unstructured and semistructured data types, such as text, audio, and video, require additional preprocessing to derive meaning and support metadata.

3. The Value and Truth of Big Data

Data has intrinsic value. But it's of no use until that value is discovered. Equally important: How truthful is your data — and how much can you rely on it?

Today, big data has become capital. Think of some of the world's biggest tech companies. A large part of the value they offer comes from their data, which they're constantly analyzing to produce more efficiency and develop new products.

Recent technological breakthroughs have exponentially reduced the cost of data storage and computing, making it easier and less expensive to store more data than ever before. With an increased volume of big data now cheaper and easier to access, you can make more accurate and precise business decisions.

Finding value in big data isn't only about analyzing it. It's an entire discovery process that requires insightful analysts, business users, and executives who ask the right questions, recognize patterns, make informed assumptions, and predict behavior.

4. Big Data Use Cases

Big data can help you address a range of business activities, from customer experience to analytics. Here are just a few.

4.1 Product Development

Companies like Netflix and Procter & Gamble (abbr. P & G) use big data to anticipate customer demand. They build predictive models for new products and services by classifying key attributes of past and current products or services and modeling the relationship between those attributes and the commercial success of the offerings. In addition, P&G uses data and analytics from focus groups, social media, test markets, and early store rollouts to plan, produce, and launch new products.

4.2 Predictive Maintenance

Factors that can predict mechanical failures may be deeply buried in structured data, such as the equipment year, make, and model of a machine, as well as in unstructured data that covers millions of log entries, sensor data, error messages, and engine temperature. By analyzing these indications of potential issues before the problems happen, organizations can deploy maintenance more cost effectively and maximize parts and equipment uptime.

4.3 Customer Experience

The race for customers is on. It is more possible now than ever before to have a clearer view of customer experience. Big data enables you to gather data from social media, web visits, call logs, and other data sources to improve the interaction experience and maximize the value delivered. You start delivering personalized offers, reduce customer churn, and handle issues proactively.

4.4 Fraud and Compliance

When it comes to security, it's not just a few rogue hackers; and you're up against entire expert teams. Security landscapes and compliance requirements are constantly evolving. Big data helps you identify patterns in data that indicate fraud and aggregate large volumes of information to make regulatory reporting much faster.

4.5 Machine Learning

Machine learning is a hot topic right now. And data—specifically big data—is one of the reasons why it is so. We are now able to teach machines instead of programing them. The availability of big data to train machine-learning models makes that happen.

4.6 Operational Efficiency

Operational efficiency may not always make the news, but it's an area in which big data is having the most impact. With big data, you can analyze and assess production, customer feedback and returns, and other factors to reduce outages and anticipate future demands. Big data can also be used to improve decision-making in line with current market demand.

4.7 Drive Innovation

Big data can help you innovate by studying interdependencies among humans, institutions, entities and process, and then determining new ways to use those insights. You can use data insights to improve decisions about financial and planning considerations, examine trends and what new products and services customers want to deliver and implement dynamic pricing. There are endless possibilities.

5. Big Data Challenges

While big data holds a lot of promise, it is not without its challenges.

First, big data is really big. Although new technologies have been developed for data storage, data volumes are doubling in size about every two years. Organizations still struggle to keep pace with their data and find ways to effectively store it.

Second, it's not enough to just store the data. Data must be used to be valuable and that depends on curation. Clean data, or data that's relevant to the client and organized in a way that enables meaningful analysis, requires a lot of work. Data scientists spend 50 to 80 percent of their time curating and preparing data before it can actually be used.

Finally, big data technology is changing at a rapid pace.

6. How Big Data Works

Big data gives you new insights that open up new opportunities and business models. Getting started involves three key actions.

6.1 Integrate

Big data brings together data from many disparate sources and applications. Traditional data integration mechanisms, such as ETL (extract, transform, and load) generally aren't up to the task. It requires new strategies and technologies to analyze big data sets at terabyte, or even petabyte, scale.

During integration, you need to bring in the data, process it, and make sure it's formatted and available in a form that your business analysts can get started with.

6.2 Manage

Big data requires storage. Your storage solution can be in the cloud, on premises, or both. You can store your data in any form you want and bring your desired processing requirements and necessary process engines to those data sets on an on-demand basis. Many people choose their storage solution according to where their data is currently residing. The cloud is gradually gaining popularity because it supports your current compute requirements and enables you to spin up resources as needed.

6.3 Analyze

Your investment in big data pays off when you analyze and act on your data. You get new clarity with a visual analysis of your varied data sets. You can explore the data further to make new discoveries and share your findings with others. You can build data models with machine learning and artificial intelligence,and put your data to work as well.

New Words

variety	[vəˈraɪətɪ]	*n.* 多样；种类；多样化
velocity	[vəˈlɒsətɪ]	*n.* 速率，速度；高速，快速
voluminous	[vəˈluːmɪnəs]	*adj.* 大的；多卷的 *adv.* 容量大地 *n.* 容量大
software	[ˈsɒftweə]	*n.* 软件
massive	[ˈmæsɪv]	*adj.* 大的；大块的，大量的；大规模的
address	[əˈdres]	*v.* 处理

		n. 地址
tackle	[ˈtækl]	*vt.* 着手处理
		n. 用具，装备
volume	[ˈvɒljuːm]	*n.* 量，大量
		adj. 大量的
matter	[ˈmætə]	*n.* 事件；（讨论、考虑等的）问题；重要性；
		vi. 要紧，重要；有重大影响；有重要性
low-density	[ˈləuˈdensɪtɪ]	*adj.* 低密度的
unstructured	[ʌnˈstrʌktʃəd]	*adj.* 无结构的，未组织的
mobile	[ˈməubaɪl]	*adj.* 可移动的
app	[æp]	*abbr.* 应用程序（application）
equipment	[ɪˈkwɪpmənt]	*n.* 设备，装备
terabyte	[ˈterəbaɪt]	*n.* 太字节（2^{40}字节）
petabyte	[ˈpetəbaɪt]	*n.* 拍字节（2^{50}字节）
rate	[reɪt]	*n.* 速度；比率；等级
		vt. 估价；值得；定级
memory	[ˈmemərɪ]	*n.* 存储器，内存
disk	[dɪsk]	*n.* 磁盘
type	[taɪp]	*n.* 类型
		vt. 按类型把……归类
traditional	[trəˈdɪʃənl]	*adj.* 传统的；惯例的
database	[ˈdeɪtəbeɪs]	*n.* 数据库
semistructured	[ˌsemɪˈstrʌktʃəd]	*adj.* 半结构化的
preprocessing	[prɪˈprəusesɪŋ]	*n.* 预处理，预加工
truthful	[ˈtruːθful]	*adj.* 说实话的，真实的
capital	[ˈkæpɪtl]	*n.* 资本；资源
analyze	[ˈænəlaɪz]	*v.* 分析
efficiency	[ɪˈfɪʃənsɪ]	*n.* 效率，效能；实力，能力
develop	[dɪˈveləp]	*vt.* 开发；研制
		vi. 发展
technological	[ˌtɛknəˈlɑdʒɪkəl]	*adj.* 技术上的
breakthrough	[ˈbrekˌθru]	*n.* 突破；重要技术成就
exponentially	[ˌekspəˈnenʃəlɪ]	*adv.* 以指数方式
storage	[ˈstɔrɪdʒ]	*n.* 存储
compute	[kəmˈpjuːt]	*v.* 计算，估算；用计算机计算
		n. 计算
precise	[prɪˈsaɪs]	*adj.* 清晰的；精确的

insightful	['ɪnˌsaɪtful]	*adj.* 富有洞察力的，有深刻见解的
predict	[prɪ'dɪkt]	*v.* 预言，预测
predictive	[prɪ'dɪktɪv]	*adj.* 预测的，预言性的
indication	[ˌɪndɪ'keɪʃn]	*n.* 指示；象征；表明；标示
maximize	['mæksɪˌmaɪz]	*vt.* 最大化
		vi. 达到最大值
uptime	['ʌptaɪm]	*n.*（计算机等的）正常运行时间
compliance	[kəm'plaɪəns]	*n.* 合规
rogue	[rəʊg]	*n.* 流氓，无赖
		vt. 欺骗
regulatory	['regjʊlətərɪ]	*adj.* 监管的，调整的
train	[treɪn]	*v.* 训练，教育，培养
impact	['ɪmpækt]	*n.* 影响
	[ɪm'pækt]	*vt.* 对……产生影响
		vi. 产生影响
assess	[ə'ses]	*vt.* 评估
feedback	['fiːdbæk]	*n.* 反馈，反应
innovation	[ˌɪnə'veɪʃn]	*n.* 改革，创新；新观念，新发明，新设施
innovate	['ɪnəveɪt]	*vi.* 改革，创新
		vt. 引入（新事物、思想或方法）
interdependency	[ɪntədɪ'pendənsɪ]	*n.* 相互依赖，相互依存；相互依赖性；依赖关系
institution	[ˌɪnstɪ'tjuːʃn]	*n.* 机构
decision	[dɪ'sɪʒən]	*n.* 决定
financial	[faɪ'nænʃəl]	*adj.* 财政的，财务的，金融的
deliver	[dɪ'lɪvə]	*vt.* 递送，交付
		vi. 投递，传送
implement	['ɪmplɪmənt]	*vt.* 实施，执行；使生效，实现
		n. 工具，器械；手段
endless	['endlɪs]	*adj.* 无尽的，无边的
possibility	[pɒsɪ'bɪlɪtɪ]	*n.* 可能，可能性
promise	['prɒmɪs]	*vt.* 许诺；给人以……的指望或希望；保证
		vi. 许诺；有指望，有前途
		n. 许诺；希望，指望
technology	[tek'nɒlədʒɪ]	*n.* 科技（总称），技术
valuable	['væljʊəbl]	*adj.* 有价值的，可评估的
strategy	['strætədʒɪ]	*n.* 策略，战略
format	['fɔːmæt]	*vt.* 使格式化

engine	[ˈɛndʒɪn]	*n.* 发动机，引擎
on-demand	[ɒndɪˈmɑːnd]	*n.* 按需
gradually	[ˈgrædʒʊəlɪ]	*adv.* 逐步地，渐渐地
popularity	[ˌpɒpjuˈlærɪtɪ]	*n.* 普及，流行
clarity	[ˈklærɪtɪ]	*n.* 清楚，明晰；透明；明确
explore	[ɪkˈsplɔː]	*v.* 探索，探究
discover	[dɪsˈkʌvə]	*vt.* 发现，获得知识

Phrases

big data	大数据
be known as	被称为
data set	数据集
data processing	数据处理
unstructured data	非结构化数据
data stream	数据流
real time	实时
relational database	关系数据库
semistructured data	半结构化数据
data storage	数据存储
customer experience	客户体验
use case	用例，使用案例，用况
customer demand	客户需求
predictive model	预测模型
focus group	焦点小组
social media	社交媒体
be buried in	埋入
error message	出错消息
customer churn	客户流失
identify pattern	识别模式
machine learning	机器学习
dynamic pricing	动态定价
a lot of	许多的，诸多
depend on	取决于
rapid pace	快节奏
business model	商业模式

get started with 开始
storage solution 存储解决方案
according to 根据，按照
spin up 启动
artificial intelligence 人工智能

Abbreviations

ETL（Extract, Transform, Load） 提取、转换和加载

Exercises

【Ex.1】Answer the following questions according to the text.

1. What is big data according to Gartner?
2. What are the three Vs of big data?
3. What does finding value in big data mean?
4. How do companies like Netflix and Procter & Gamble build predictive models for new products and services?
5. How can organizations deploy maintenance more cost effectively and maximize parts and equipment uptime?
6. What is an area in which big data is having the most impact? What can you do with big data?
7. How can big data help you innovate?
8. What are the challenges big data has?
9. What new insights does big data give you? What are the three key actions getting started involves?
10. Why is the cloud gradually gaining popularity?

【Ex.2】Translate the following terms or phrases from English into Chinese and vice versa.

1. data processing	1.
2. data storage	2.
3. data set	3.
4. relational database	4.
5. machine learning	5.
6. *n.* 数据库	6.
7. *n.* 预处理，预加工	7.
8. 预测模型	8.

9. *adj.* 易访问的，可存储的 ________ 9. ________

10. *n.* 发动机，引擎 ________ 10. ________

【Ex.3】Translate the following passages into Chinese.

Big Data as a Service

Big data as a service (BDaaS) is the delivery of statistical analysis tools or information by an outside provider that helps organizations understand and use insights gained from large information sets in order to gain a competitive advantage.

Given the immense amount of unstructured data generated on a regular basis, BDaaS is intended to free up organizational resources by taking advantage of the predictive analytics skills of an outside provider to manage and assess large data sets, rather than hiring in-house staff for those functions. It can take the form of software that assists with data processing or a contract for the services of a team of data scientists.

BDaaS is a form of managed services, similar to Software as a Service or Infrastructure as a Service. It often relies upon cloud storage to preserve continual data access for the organization that owns the information as well as the provider working with it.

【Ex.4】Fill in the blanks with the words given below.

fraud	impact	critical	market	patient
manage	insights	privacy	accurately	preventing

Who Uses Big Data

Big data affects organizations across practically every industry. See how each industry can benefit from this onslaught of information.

1. Banking

With large amounts of information streaming in from countless sources, banks are faced with finding new and innovative ways to ____1____ big data. While it's important to understand customers and boost their satisfaction, it's equally important to minimize risk and ____2____ while maintaining regulatory compliance. Big data brings big ____3____, but it also requires financial institutions to stay one step ahead of the game with advanced analytics.

2. Education

Educators armed with data-driven insight can make a significant ____4____ on school systems, students and curriculums. By analyzing big data, they can identify at-risk students, make sure students are making adequate progress, and can implement a better system for evaluation and support of teachers and principals.

3. Government

When government agencies are able to harness and apply analytics to their big data, they gain significant ground when it comes to managing utilities, running agencies, dealing with traffic congestion or ___5___ crime. While there are many advantages to big data, governments must also address issues of transparency and ___6___.

4. Health Care

Patient records. Treatment plans. Prescription information. When it comes to health care, everything needs to be done quickly, ___7___ — and, in some cases, with enough transparency to satisfy stringent industry regulations. When big data is managed effectively, health care providers can uncover hidden insights that improve ___8___ care.

5. Manufacturing

Armed with insight that big data can provide, manufacturers can boost quality and output while minimizing waste — processes that are key in today's highly competitive ___9___. More and more manufacturers are working in an analytics-based culture, which means they can solve problems faster and make more agile business decisions.

6. Retail

Customer relationship building is ___10___ to the retail industry, and the best way to manage that is to manage big data. Retailers need to know the best way to market to customers, the most effective way to handle transactions, and the most strategic way to bring back lapsed business. Big data remains at the heart of all those things.

扫码听课文

Text B
Big Data Analytics

Big data analytics is quickly gaining adoption. Enterprises have awakened to the reality that their big data stores represent a largely untapped gold mine that could help them lower costs, increase revenues and become more competitive. They don't just want to store their vast quantities of data, they want to convert that data into valuable insights that can help improve their companies.

As a result, investment in big data analytics tools is seeing remarkable gains. According to IDC, worldwide sales of big data and business analytics tools are likely to reach $150.8 billion in 2017, which is 12.4 percent higher than that in 2016. And the market research firm doesn't see that trend stopping anytime soon. It forecasts 11.9 percent annual growth through 2020 when revenues will top $210 billion.

Clearly, the trend toward big data analytics is here to stay. IT professionals need to familiarize themselves with the topic if they want to remain relevant within their companies.

1. What is Big Data Analytics

The term " big data" refers to digital stores of information that have a high volume, velocity and variety. Big data analytics is the process of using software to uncover trends, patterns, correlations or other useful insights in those large stores of data.

Data analytics isn't new. It has been around for decades in the form of business intelligence and data mining software. Over the years, that software has improved dramatically so that it can handle much larger data volumes, run queries more quickly and perform more advanced algorithms.

The market research firm Gartner categorizes big data analytics tools into four different categories.

- Descriptive Analytics: These tools tell companies what happened. They create simple reports and visualizations that show what occurred at a particular point in time or over a period of time. These are the least advanced analytics tools.
- Diagnostic Analytics: Diagnostic tools explain why something happened. More advanced than descriptive reporting tools, they allow analysts to dive deep into the data and determine root causes for a given situation.
- Predictive Analytics: Among the most popular big data analytics tools available today, predictive analytics tools use highly advanced algorithms to forecast what might happen next. Often these tools make use of artificial intelligence and machine learning technology.
- Prescriptive Analytics: A step above predictive analytics, prescriptive analytics tools tell organizations what they should do in order to achieve a desired result. These tools require very advanced machine learning capabilities, and few solutions on the market today offer true prescriptive capabilities.

2. Benefits of Big Data Analytics

Organizations decide to deploy big data analytics for a wide variety of reasons, including the following.

2.1 Business Transformation

In general, executives believe that big data analytics offers tremendous potential for revolutionizing their organizations. In the 2016 Data & Analytics Survey from IDGE, 78 percent of people surveyed agreed that over the next one to three years the collection and analysis of big data could fundamentally change the way their companies do business.

2.2 Competitive Advantage

In the MIT Sloan Management Review Research Report Analytics as a Source of Business Innovation, sponsored by SAS, 57 percent of enterprises surveyed said their use of analytics was helping them achieve competitive advantage, up from 51 percent who said the same thing in 2015.

2.3 Innovation

Big data analytics can help companies develop products and services that appeal to their customers and identify new opportunities for revenue generation. Also in the MIT Sloan Management survey, 68 percent of respondents agreed that analytics has helped their company innovate. That's an increase from 52 percent in 2015.

2.4 Lower Costs

In the NewVantage Partners Big Data Executive Survey 2017, 49.2 percent of companies surveyed said that they had successfully decreased expenses as a result of a big data project.

2.5 Improved Customer Service

Organizations often use big data analytics to examine social media, customer service, sales, and marketing data. This can help them better gauge customer sentiment and respond to customers in real time.

2.6 Increased Security

Another key area for big data analytics is IT security. Security software creates an enormous amount of log data. By applying big data analytics techniques to this data, organizations can sometimes identify and thwart cyberattacks that would otherwise have gone unnoticed.

3. Big Data Analytics Challenges

Implementing a big data analytics solution isn't always as straightforward as companies hope it will be. In fact, most surveys find that the number of organizations experiencing a measurable financial benefit from their big data analytics lags behind the number of organizations implementing big data analytics. Several different obstacles can make it difficult to achieve the benefits promised by big data analytics vendors.

3.1 Data Growth

One of the biggest challenges of big data analytics is the explosive rate of data growth. According to IDC, the amount of data in the world's servers is roughly doubling every two years. By 2020, those servers will likely hold 44 zettabytes of digital information. To put that in perspective, that is enough data to fill a stack of iPads stretching from the earth to the moon 6.6

times. Big data analytics solutions must be able to perform well at scale if they are going to be useful to enterprises.

3.2 Unstructured Data

Most of the data stored in an enterprise's systems doesn't reside in structured databases. Instead, it is unstructured data, such as email messages, images, reports, audio files, videos, and other types of files. This unstructured data can be very difficult to search—unless you have advanced artificial intelligence capabilities. Vendors are constantly updating their big data analytics tools to make them better at examining and extracting insights from unstructured data.

3.3 Data Siloes

Enterprise data is created by a wide variety of different applications, such as enterprise resource planning (ERP) solutions, customer relationship management (CRM) solutions, supply chain management software, ecommerce solutions, office productivity programs, etc. Integrating the data from all these different sources is one of the most difficult challenges in any big data analytics project.

3.4 Cultural Challenges

Although big data analytics is becoming commonplace, it hasn't infiltrated the corporate culture everywhere yet. In the NewVantage Partners Survey, 52.5 percent of executives said that organizational hurdles like lack of alignment, internal resistance or lack of coherent strategy were preventing them from using big data as widely as they would have liked.

4. Big Data Analytics Trends

What's coming next for the big data analytics market? Experts offer a number of predictions.

4.1 Open Source

As big data analytics increases its momentum, the focus is on open-source tools that help break down and analyze data. Hadoop, Spark and NoSQL databases are the winners here. Even proprietary tools now incorporate leading open source technologies and/or support those technologies. That seems unlikely to change for the foreseeable future.

4.2 Market Segmentation

Plenty of general-purpose big data analytics platforms have hit the market, but expect even more to emerge that focus on specific niches, such as security, marketing, CRM, application performance monitoring and hiring. Analytics tools are also being integrated into existing enterprise software at a rapid rate.

4.3 Artificial Intelligence and Machine Learning

As interest in AI has skyrocketed, vendors have rushed to incorporate machine learning and cognitive capabilities into their big data analytics tools. According to Gartner, by 2020, almost every new software product, including big data analytics, will incorporate AI technologies. In addition, the company said, “By 2020, AI will be a top five investment priority for more than 30 percent of CIOs.”

4.4 Prescriptive Analytics

Fueled by this rush to AI, companies are expected to become more interested in prescriptive analytics. Seen by many as the “ultimate” type of big data analytics, these tools will not only be able to predict the future, but also to suggest courses of action that might lead to desirable results for organizations. But before these types of solutions can become mainstream, vendors will need to make advancements in both hardware and software.

4.5 Refocusing on the Human Decision-Making

As machine learning improves and becomes a table stakes feature in analytics suites, don’t be surprised if the human element initially gets downplayed, before coming back into vogue.

Two of the most famous big data prognosticators/pioneers are Billy Beane and Nate Silver. Beane popularized the idea of correlating various statistics with under-valued player traits in order to field an A’s baseball team on the cheap that could compete with deep-pocketed teams like the Yankees.

Meanwhile, Nate Silver’s effect was so strong that people who didn’t want to believe his predictions created all sorts of analysis-free zones, such as Unskewed Polls (which, ironically, were ridiculously skewed). Many think of Silver as a polling expert, but Silver is also a master at big data analysis.

In each case, what mattered most was not the machinery that gathered in the data and formed the initial analysis, but the human on top analyzing what this all meant. People can look at polling data and pretty much treat them as Rorscharch tests. Silver, on the other hand, pours over reams of data, looks at how various polls have performed historically, factors in things that could influence the margin of error (such as the fact that younger voters are often under-counted since they don’t have landline phones) and emerges with incredibly accurate predictions.

Similarly, every baseball GM now values on-base percentage and other advanced stats, but few are able to compete as consistently on as little money as Beane’s A’s teams can. There’s more to finding under-valued players than crunching numbers. You also need to know how to push the right buttons in order to negotiate trades with other GMs, and you need to find players who will fit into your system.

As big data analytics becomes mainstream, it will be like many earlier technologies. Big Data analytics will be just another tool. What you do with it, though, will be what matters.

New Words

adoption	[ə'dɒpʃn]	*n.* 采用
awaken	[ə'weɪkən]	*vt.& vi.* 唤醒，觉醒；（使）意识到
untapped	[ˌʌn'tæpt]	*adj.* 未开发的，未利用的
competitive	[kəm'petɪtɪv]	*adj.* 竞争的，比赛的；（价格等）有竞争力的；（指人）好竞争的
vast	[vɑːst]	*adj.* 巨大的，大量的
quantity	['kwɒntɪtɪ]	*n.* 量，数量；定量，大批
firm	[fɜːm]	*n.* 商号，商行；公司，企业 *vt. & vi.* 使坚固，使坚实
professional	[prə'feʃənl]	*n.* 专业人士，专家 *adj.* 专业的，职业的
familiarize	[fə'mɪlɪəraɪz]	*vt.* 使（某人）熟悉，使通晓
algorithm	['ælgərɪðəm]	*n.* 算法
category	['kætəgərɪ]	*n.* 种类，类别
descriptive	[dɪ'skrɪptɪv]	*adj.* 描述的
diagnostic	[ˌdaɪəg'nɒstɪk]	*adj.* 诊断的，判断的；特征的 *n.* 诊断法，诊断程序
explain	[ɪk'spleɪn]	*vt. & vi.* 说明，解释，讲解
situation	[ˌsɪtjʊ'eɪʃn]	*n.* 情况，局面，形势，处境
prescriptive	[prɪ'skrɪptɪv]	*adj.* 规定的，指定的，惯例的
achieve	[ə'tʃiːv]	*vt.* 取得，实现，成功 *vi.* 达到预期的目的，实现预期的结果
capability	[ˌkeɪpə'bɪlɪtɪ]	*n.* 能力；容量；性能
deploy	[dɪ'plɔɪ]	*vt. & vi.* 使展开；施展；有效地利用
executive	[ɪg'zekjʊtɪv]	*n.* 总经理；行政部门 *adj.* 执行的；管理的
fundamentally	[ˌfʌndə'mentəlɪ]	*adv.* 根本地
appeal	[ə'piːl]	*vi.* 有吸引力
examine	[ɪg'zæmɪn]	*v.* 检查，调查
gauge	[geɪdʒ]	*n.* 评估 *vt.* 评估，判断
sentiment	['sentɪmənt]	*n.* 意见，观点
thwart	[θwɔːt]	*vt.* 阻挠，挫败
cyberattack	['saibərəˌtæk]	*n.* 网络攻击

unnoticed [ˌʌn'nəʊtɪst] *adj.* 未觉察的，未注意的；被忽视的

measurable ['meʒərəbl] *adj.* 可测量的，可预见的

benefit ['benɪfɪt] *n.* 利益，好处
vt. 有益于，有助于；得益，受益

roughly ['rʌflɪ] *adv.* 粗略地，大体上，大致上

zettabyte ['zetəbaɪt] *n.* 泽字节（2^{70}字节）

stretch [stretʃ] *v.* 伸展；延伸
n. 伸展；弹性
adj. 可伸缩的；弹性的

extract [ɪk'strækt] *v.* 提取，提炼，选取

ecommerce [iː'kɒmɜːs] *n.* 电子商务

productivity [ˌprɒdʌk'tɪvɪtɪ] *n.* 生产率，生产力

commonplace ['kɒmənpleɪs] *adj.* 平凡的，普通的
n. 寻常的事物

infiltrate ['ɪnfɪltreɪt] *vt. & vi.*（使）渗透，（使）渗入；（使）潜入

hurdle ['hɜːdl] *n.* 障碍，困难
v. 克服困难

alignment [ə'laɪnmənt] *n.* 队列，排成直线；校直，调整

resistance [rɪ'zɪstəns] *n.* 阻力

trend [trend] *n. & vi.* 倾向，趋势

expert ['ekspɜːt] *n.* 专家，权威，行家，高手
adj. 专家的

momentum [mə'mentəm] *n.* 势头；动力

proprietary [prə'praɪətrɪ] *adj.* 专有的，专利的；（商品）专卖的
n. 所有权，所有物

incorporate [ɪn'kɔːpəreɪt] *vt.* 使混合，协作
vi. 包含；吸收；合并；混合

foreseeable [fɔː'siːəbl] *adj.* 可预见到的

plenty ['plentɪ] *n.* 丰富，大量；充足，充分
adv. 相当地，充分地
adj. 足够的，很多的，充裕的

emerge [i'mɜːdʒ] *vi.* 出现，浮现

niche [nɪtʃ] *n.* 有利可图的缺口，商机

skyrocket ['skaɪrɒkɪt] *vi.* 突升，猛涨

rush [rʌʃ] *vi.*（使）急速行进，仓促完成；猛攻
vt.（使）仓促行事
n. 匆忙

fuel	[ˈfjuːəl]	*n.* 燃料 *vt.* 给……加燃料，给……加油；激起 *vi.* 补充燃料
ultimate	[ˈʌltɪmɪt]	*adj.* 最后的；极限的；首要的；最大的 *n.* 终极，顶点
advancement	[ədˈvɑːnsmənt]	*n.* 前进，进步；提升，升级
decision-making	[dɪˈsɪʒnˈmeɪkɪŋ]	*n.* 决策 *adj.* 决策的
suite	[swiːt]	*n.* 套件
downplay	[ˌdaʊnˈpleɪ]	*vt.* 贬低，轻视，不予重视
vogue	[vəʊg]	*n.* 时尚，流行；时髦的事物 *adj.* 流行的，时髦的
famous	[ˈfeɪməs]	*adj.* 著名的，出名的
prognosticator	[prɒgˈnɒstɪkeɪtə]	*n.* 预言者
pioneer	[ˌpaɪəˈnɪə]	*n.* 先驱者，创始者 *vt.* 开拓，开发；做（……的）先锋；提倡
deep-pocketed	[diːpˈpɔkɪtɪd]	*adj.* 有钱的，财大气粗的
ridiculously	[rɪˈdɪkjuləslɪ]	*adv.* 可笑地，荒谬地
initial	[ɪˈnɪʃl]	*adj.* 最初的，开始的
influence	[ˈɪnfluəns]	*n.* 影响；势力；有影响的人（或事物） *vt.* 影响；支配；对……起作用
incredibly	[ɪnˈkredəblɪ]	*adv.* 难以置信地；很，极为
negotiate	[nɪgəʊʃ ɪeɪt]	*vi.* 谈判，协商，交涉 *vt.* 谈判达成；议价出售
mainstream	[ˈmeɪnstriːm]	*n.* 主流；主要倾向，主要趋势

Phrases

analytics tool	分析工具
data mining	数据挖掘
market research	市场调研，市场调查
a period of	一段时间
diagnostic tool	诊断工具
predictive analytics tool	预测分析工具
prescriptive analytics	规范性分析，常规分析
business transformation	企业转型，业务转型

as well as	也，又
log data	日志数据，记录数据
a stack of	一摞，一堆
at scale	在一定范围内
audio file	音频文件
supply chain management	供应链管理
open-source tool	开源工具
break down	划分
market segmentation	市场细分
be integrated into ...	被集成到……
a table stake	桌面筹码，赌注
on the cheap	便宜地
pretty much	几乎，差不多
landline phone	座机，固定电话

Abbreviations

MIT（Massachusetts Institute of Technology）	麻省理工学院
IT（Information Technology）	信息技术
IDC（International Data Corporation）	国际数据公司
ERP（Enterprise Resource Planning）	企业资源计划
CRM（Customer Relationship Management）	客户关系管理
AI（Artificial Intelligence）	人工智能
CIO（Chief Information Officer）	首席信息官

Exercises

【Ex.5】Answer the following questions according to the text.

1. What reality have enterprises awakened to?
2. What does the term “big data” refer to? What is big data analytics?
3. How many different categories does the market research firm Gartner categorize big data analytics tools into? What are they?
4. What can big data analytics help companies?
5. What does security software create? What can organizations do by applying big data analytics techniques to it?
6. What are the different obstacles which can make it difficult to achieve the benefits promised by big

data analytics vendors?

7. How is enterprise data created?

8. What is the focus as big data analytics increases its momentum?

9. Fueled by this rush to AI, what are companies expected to do? Why?

10. Why did Beane popularize the idea of correlating various statistics with under-valued player traits?

参考译文

什么是大数据

1. 大数据的定义

大数据是包含更多种类的数据，其数量越来越多，获取速度也越来越快，这个定义来自加特纳，被称为3V。

简而言之，大数据是更大、更复杂的数据集，尤其是来自新数据源的。这些数据集非常庞大，以至于传统的数据处理软件无法管理它们。但是，这些海量数据可用于解决以前无法解决的业务问题。

2. 大数据的3V

2.1 大量

数据量很重要。大数据，意味着你必须处理大量低密度、非结构化的数据。这可以是未知价值的数据，如Twitter反馈的数据，网页或移动应用上的点击流，或来自有效传感器设备的数据。对有些组织来说，这可能是数十TB的数据，而对其他组织来说，可能是数百PB的数据。

2.2 高速

高速是指接收和应对数据的高速率。一些支持互联网的智能产品实时或接近实时运行，需要实时评估和应对。

2.3 多样

多样是指可用的数据种类繁多。传统数据类型是结构化的，特别适合关系数据库。随着大数据的兴起，大量新的非结构化数据出现。非结构化和半结构化数据类型，如文本、音频和视频，需要额外的预处理才能获取意义和支持元数据。

3. 大数据的价值与真实性

数据具有内在价值，但是只有发现了它的价值才有用。同样重要的是：数据的真实性如何，即可依赖的程度如何？

今天，大数据已成为资本。想想世界上一些大型科技公司，它们提供的大部分价值来自于它们的数据。它们不断分析这些数据以提高效率并开发新产品。

最近的技术突破极大地降低了数据存储和计算的成本，使存储更多的数据比以往更容易、成本更低。随着现在更便宜、更易于访问的大数据量的增加，你可以做出更精准的业务决策。

在大数据中寻找价值不仅仅是对它进行分析。这是一个完整的发现过程，需要富有洞察力的分析师、业务用户和高管，他们会提出正确的问题、识别模式，做出切合实际的假设并预测行为。

4. 大数据实用范例

大数据可以帮助处理从客户体验到分析的各种业务活动。以下是一些范例。

4.1 产品开发

Netflix 和 Procter&Gamble 等公司使用大数据预测客户需求。它们通过对过去和当前产品或服务的关键属性进行分类，并建模分析这些属性与产品的商业成功之间的关系，构建了新产品和服务的预测模型。此外，宝洁还使用焦点小组、社交媒体、测试市场和早期商店的数据和分析规划、生产和推出新产品。

4.2 预测性维护

可以预测机械故障的因素可能深深地隐藏在结构化数据中，如设备年份、品牌和机器型号以及数百万个日志条目、传感器数据、错误信息和发动机温度等的非结构化数据。通过在问题发生之前分析这些潜在问题的迹象，组织可以更经济地部署维护并尽量延长部件和设备的正常运行时间。

4.3 客户体验

争夺客户无时不在。现在比以往能更加清晰地了解客户体验。通过大数据，你可以从社交媒体、网络访问、呼叫日志和其他数据源收集数据，从而改善交互体验并最大限度地提高交付价值。开始提供个性化优惠，减少客户流失，并主动处理问题。

4.4 欺诈和合规

当谈到安全性时，不仅是指对抗一些流氓黑客，还要对抗整个专家团队。安全环境和合规性要求不断变化。大数据有助于识别数据中显示欺诈的模式并汇总大量信息，从而更快地提交监管报告。

4.5 机器学习

机器学习是当前的热门话题，而数据（特别是大数据）是其原因之一。我们现在能够教授机器而不用给它编程，这就是用大数据来训练机器学习模型实现的。

4.6 运营效率

运营效率可能并不总能创造新闻，但它是受大数据影响很大的一个领域。利用大数据可以分析和评估生产、客户反馈和退货以及其他因素，以减少停机并预测未来需求。大数据也可用于改善符合当前市场需求的决策。

4.7 推动创新

通过研究人、机构、实体和流程之间的相互依赖关系并找到利用这些关系的新方式，大数据能够帮助人们进行创新。你还可以利用数据洞察力改进财务和计划的决策，验证趋势及客户希望提供的新产品和服务，并实施动态定价。这有无穷无尽的可能性。

5. 大数据面临的挑战

虽然大数据具有很好的前景，但它并非没有挑战。

首先，大数据实在大。尽管已经开发了用于数据存储的新技术，但是大约每两年数据量就增加一倍。组织仍在努力应对源源不断的数据并找到有效存储数据的方法。

其次，仅存储数据还不够。数据必须得到使用才有价值，这取决于策略。清洁数据或按照有意义分析的方式组织客户数据，需要大量工作。数据科学家花费 50%～80%的时间在实际使用数据之前策划和准备数据。

最后，大数据技术正在快速变化。

6. 大数据如何工作

大数据为你提供新的见解，开辟新的机会和商业模式。其入门涉及三个关键操作。

6.1 整合

大数据汇集了许多不同来源和应用程序的数据。传统的数据集成机制，例如 ETL（提取，转换和加载）通常不能胜任该任务。它需要新的策略和技术来分析 TB 级甚至 PB 级的大数据集。在集成过程中，需要引入数据，处理数据，并确保以业务分析师可以使用的格式提供数据。

6.2 管理

大数据需要存储。存储解决方案可以位于云端、内部或两者结合使用。你能够以任何所需的形式存储数据，并根据需要将所需的处理要求和必要的流程引擎提供给这些数据集。许多人根据某数据当前所在的位置选择存储方案。云存储正在逐渐普及，因为它支持当前的计算需求，并能够根据需要启动资源。

6.3 分析

在分析和处理数据时，对大数据的投资会得到回报。通过对各种数据集的可视化分析可以获得新的理解。对数据进一步探索可以获得新的发现，并与他人分享此发现。也可使用机器学习和人工智能构建数据模型，让数据更好地发挥作用。

Unit 2

Text A
Data Model (1)

1. Overview

A data model is an abstract model that organizes elements of data and standardizes how they relate to one another and to properties of the real world entities. For instance, a data model may specify that the data element representing a car be composed of a number of other elements which, in turn, represent the color and size of the car and define its owner.

The term data model is used in two distinct but closely related senses. Sometimes it refers to an abstract formalization of the objects and relationships found in a particular application domain, for example the customers, products, and orders found in a manufacturing organization. At other times it refers to a set of concepts used in defining such formalizations: For example concepts such as entities, attributes, relations, or tables. So the "data model" of a banking application may be defined using the entity-relationship "data model". This article uses the term in both senses.

Data model is based on data, data relationship, data semantic and data constraint. A data model provides the details of information to be stored.

A data model can sometimes be referred to as a data structure, especially in the context of programming languages. Data models are often complemented by function models, especially in the context of enterprise models.

The main aim of data models is to support the development of information systems by providing the definition and format of data.

A data model explicitly determines the structure of data. Typical applications of data models include database models, design of information systems, and enabling exchange of data. Usually data models are specified in a data modeling language.

2. Three Perspectives of Data Model

A data model instance may be one of three kinds (See Figure 2-1) according to ANSI in 1975.

(1) Conceptual data model: It describes the semantics of a domain, being the scope of the model. For example, it may be a model of the interest area of an organization or industry. This consists of entity classes, representing kinds of things of significance in the domain, and relationship assertions about associations between pairs of entity classes. A conceptual schema specifies the kinds of facts or propositions that can be expressed using the model. In that sense, it defines the allowed expressions in an artificial "language" with a scope that is limited by the scope of the model.

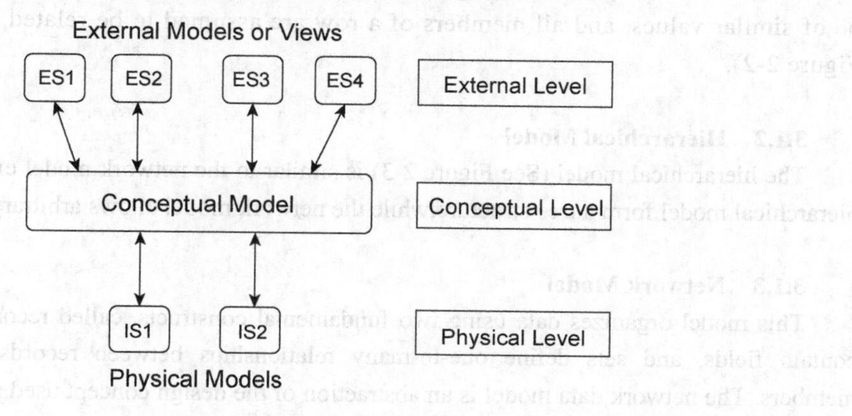

Figure 2-1 The ANSI/SPARC Three Level Architecture

(2) Logical data model: It describes the semantics, as represented by a particular data manipulation technology. This consists of descriptions of tables and columns, object oriented classes, and XML tags, among other things.

(3) Physical data model: It describes the physical means by which data are stored. This is concerned with partitions, CPUs, tablespaces, and the like.

The significance of this approach, according to ANSI, is that it allows the three perspectives to be relatively independent of each other. Storage technology can change without affecting either the logical or the conceptual model. The table/column structure can change without (necessarily) affecting the conceptual model. In each case, of course, the structures must remain consistent with the other model. The table/column structure may be different from a direct translation of the entity classes and attributes, but it must ultimately carry out the objectives of the conceptual entity class structure. Early phases of many software development projects emphasize the design of a conceptual data model. Such a design can be detailed into a logical data model. In later stages, this model may be translated into physical data model. However, it is also possible to implement a conceptual model directly.

3. Types of Data Models

3.1 Database Model

A database model is a specification describing how a database is structured and used. Several such models have been suggested. Common models are as follows.

3.1.1 Flat Model

This may not strictly qualify as a data model. The flat (or table) model consists of a single, two-dimensional array of data elements, where all members of a given column are assumed to be of similar values, and all members of a row are assumed to be related to one another (See Figure 2-2).

3.1.2 Hierarchical Model

The hierarchical model (See Figure 2-3) is similar to the network model except that links in the hierarchical model form a tree structure,while the network model allows arbitrary graph.

3.1.3 Network Model

This model organizes data using two fundamental constructs, called records and sets. Records contain fields, and sets define one-to-many relationships between records: One owner, many members. The network data model is an abstraction of the design concept used in the implementation of databases (See Figure 2-4).

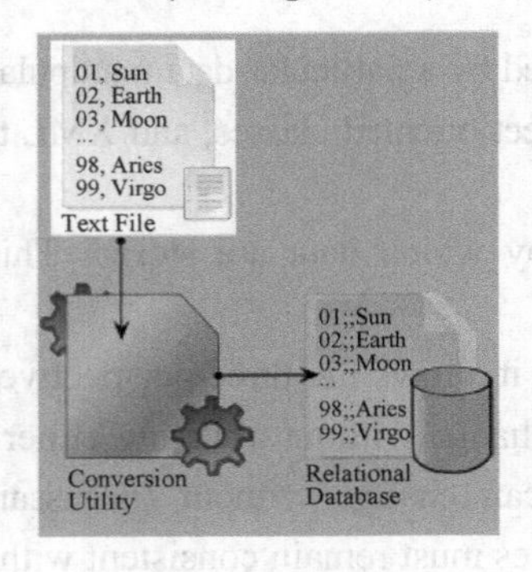

Figure 2-2 Flat Model

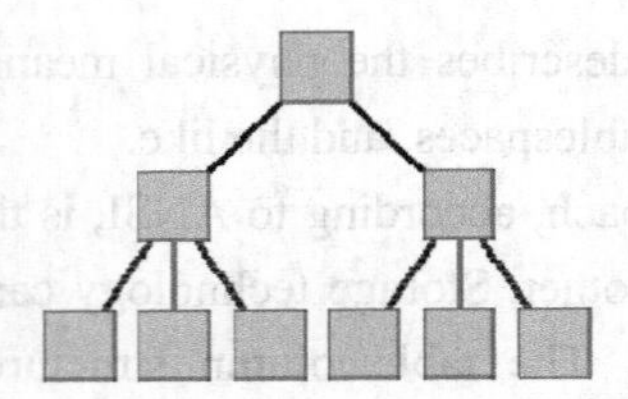

Figure 2-3 Hierarchical Model

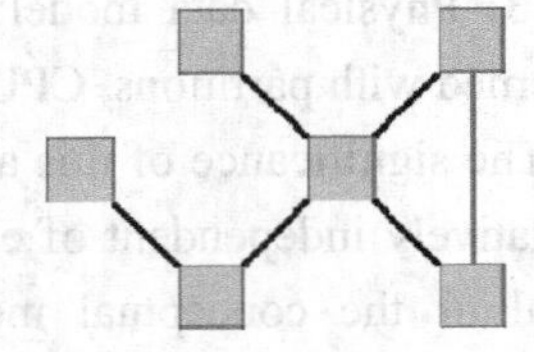

Figure 2-4 Network Model

3.1.4 Relational Model

Relational model is a database model based on first-order predicate logic. Its core idea is to describe a database as a collection of predicates over a finite set of predicate variables, describing constraints on the possible values and combinations of values.The power of the relational data model lies in its mathematical foundations and a simple user-level paradigm (See Figure 2-5).

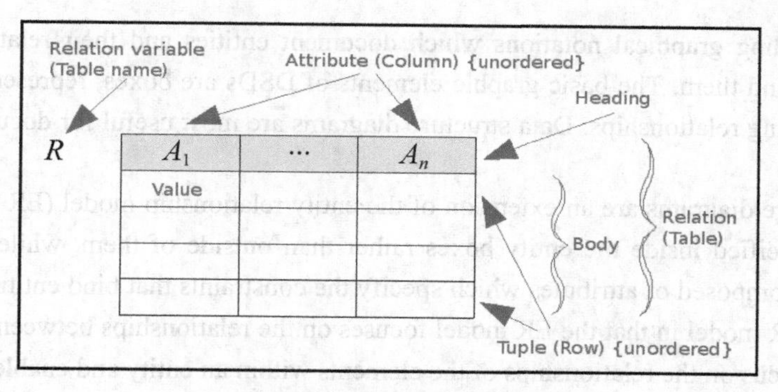

Figure 2-5 Relational Model

3.1.5 Object-Relational Model

Similar to a relational database model, only that objects, classes and inheritance are directly supported in database schemas and in the query language.

3.1.6 Object-Role Modeling

This is a method of data modeling that has been defined as "attribute free", and "fact based". The result is a verifiably correct system, from which other common artifacts, such as ERD, UML, and semantic models may be derived. Associations between data objects are described during the database design procedure, such that normalization is an inevitable result of the process.

3.1.7 Star Schema

It is the simplest style of data warehouse schema. The star schema (See Figure 2-6) consists of a few "fact tables" referencing any number of "dimension tables".

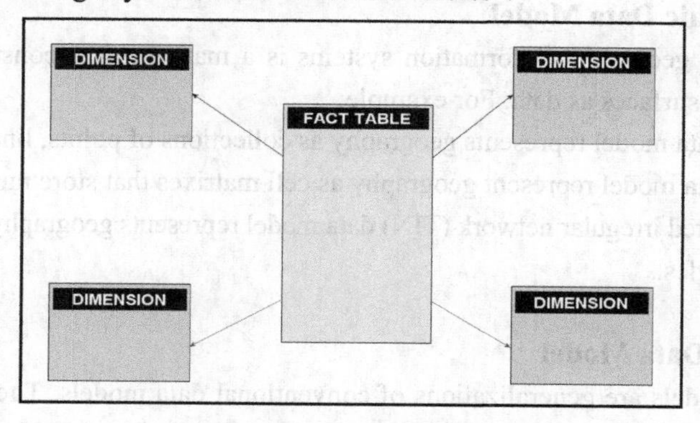

Figure 2-6 Star Schema

3.2 Data Structure Diagram

A Data Structure Diagram (DSD) is a diagram and data model used to describe conceptual data models by providing graphical notations which document entities and their relationships, and the constraints that bind them. The basic graphic elements of DSDs are boxes, representing entities, and arrows, representing relationships. Data structure diagrams are most useful for documenting complex data entities.

Data structure diagrams are an extension of the entity relationship model (ER model). In DSDs, attributes are specified inside the entity boxes rather than outside of them, while relationships are drawn as boxes composed of attributes which specify the constraints that bind entities together. DSDs differ from the ER model in that the ER model focuses on the relationships between different entities, whereas DSDs focus on the relationships of the elements within an entity and enable users to fully see the links and relationships between each entity.

There are several styles for representing data structure diagrams, with the notable difference in the manner of defining cardinality. The choices are between arrow heads, inverted arrow heads , or numerical representation of the cardinality.

3.3 Entity Relationship Model

An Entity Relationship Model (ERM), sometimes is referred to as an Entity Relationship Diagram (ERD), could be used to represent an abstract conceptual data model (or semantic data model or physical data model) used in software engineering to represent structured data. There are several notations used for ERMs. Like DSD's, attributes are specified inside the entity boxes rather than outside of them, while relationships are drawn as lines, with the relationship constraints as descriptions on the line. The ER model, while robust, can become visually cumbersome when representing entities with several attributes.

3.4 Geographic Data Model

A data model in geographic information systems is a mathematical construct for representing geographic objects or surfaces as data. For example:

(1) The vector data model represents geography as collections of points, lines, and polygons;

(2) The raster data model represent geography as cell matrixes that store numeric values;

(3) The triangulated irregular network (TIN) data model represents geography as sets of contiguous, nonoverlapping triangles.

3.5 Generic Data Model

Generic data models are generalizations of conventional data models. They define standardised general relation types, together with the kinds of things that may be related by such a relation type. Generic data models are developed as an approach to solve some shortcomings of conventional data models. For example, different modelers usually produce different conventional data models of the same domain. This can lead to difficulty in bringing the models of different people together and is an obstacle for data exchange and data integration. Invariably, however, this difference is attributable to different levels of abstraction in the models and differences in the kinds of facts that can be instantiated. The modelers need to communicate and agree on certain elements which are to be rendered more concretely, in order to make the differences less significant.

3.6 Semantic Data Model

A semantic data model is an abstraction which defines how the stored symbols relate to the real world. It is sometimes called a conceptual data model. A semantic data model in software engineering is a technique to define the meaning of data within the context of its interrelationships with other data.

The logical data structure of a database management system (DBMS), whether hierarchical, network, or relational, cannot totally satisfy the requirements for a conceptual definition of data because it is limited in scope and biased toward the implementation strategy employed by the DBMS. Therefore, the need to define data from a conceptual view has led to the development of semantic data modeling techniques. That is, techniques to define the meaning of data within the context of its interrelationships with other data. The real world, in terms of resources, ideas, events, etc., are symbolically defined within physical data stores. A semantic data model is an abstraction which defines how the stored symbols relate to the real world. Thus, the model must be a true representation of the real world.

New Words

model	['mɒdl]	*n.* 模型，模式 *vt.* 模仿；制作模型
standardize	['stændədaɪz]	*vt.* 使标准化；用标准校检
sense	[sens]	*n.* 感觉，意识，观念 *vt.* 感到；理解，领会
formalization	[ˌfɔːməlaɪ'zeɪʃn]	*n.* 形式化
manufacturing	[ˌmænjʊ'fæktʃərɪŋ]	*n.* 制造业，工业 *adj.* 制造业的，制造的
table	['teɪbl]	*n.* 表 *vt.* 制表
detail	['diːteɪl]	*n.* 细节，详述 *vt.* 详述，清晰地说明
design	[dɪ'zaɪn]	*vt.&vi.* 设计
enable	[ɪ'neɪbl]	*vt.* 使能够，使可能
exchange	[ɪks'tʃeɪndʒ]	*n.* 交换 *vt.* 交换，互换，调换

conceptual	[kən'septjuəl]	*adj.* 概念的
domain	[də'meɪn]	*n.* 范围，域
significance	[sɪg'nɪfɪkəns]	*n.* 意义，意思
assertion	[ə'sɜːʃn]	*n.* 断言；声明
schema	['skiːmə]	*n.* 模式；概要；计划
expression	[ɪk'spreʃn]	*n.* 表示，表达；表达式
manipulation	[məˌnɪpjʊ'leɪʃn]	*n.* 操作
remain	[rɪ'meɪn]	*n.* 剩余物
		vi. 留下；保持
ultimately	['ʌltɪmɪtlɪ]	*adv.* 最后，最终；根本
project	['prɒdʒekt]	*n.* 生产（或研究等）项目；方案；工程
		v. 规划，计划，拟订方案
specification	[ˌspesɪfɪ'keɪʃn]	*n.* 规范，规格；详述；说明书
array	[ə'reɪ]	*n.* 数组
set	[set]	*n.* 集合
		vt. 放置，安置；设置
field	[fiːld]	*n.* 字段
abstraction	[æb'strækʃn]	*n.* 抽象，抽象概念，抽象化
implementation	[ˌɪmplɪmen'teɪʃn]	*n.* 执行，履行；实施；完成；安装启用
predicate	['predɪkət]	*n.* 谓语，谓词
		adj. 谓语的，谓词的
inheritance	[ɪn'herɪtəns]	*n.* 继承，遗传
diagram	['daɪəgræm]	*n.* 图表；示意图
		vt. 用图表示；图解
graphical	['græfɪkl]	*adj.* 图画的，绘画的
notation	[nəʊ'teɪʃn]	*n.* 记号，标记法
document	['dɒkjʊmənt]	*n.*（计算机）文档
		vt. 证明；记录；为……提供证明
bind	[baɪnd]	*vt.* 绑定；约束；捆绑
arrow	['ærəʊ]	*n.* 箭头记号
extension	[ɪk'stenʃn]	*n.* 伸展，扩大，延长
notable	['nəʊtəbl]	*adj.* 值得注意的；显著的
cardinality	[kɑːdɪ'nælɪtɪ]	*n.* 基数
robust	[rəʊ'bʌst]	*adj.* 健壮的，强健的，结实的
cumbersome	['kʌmbəsəm]	*adj.* 麻烦的；累赘的；复杂的
vector	['vektə]	*n.* 矢量
polygon	['pɒlɪgən]	*n.* 多边形，多角形

raster	[ˈræstə]	*n.* 光栅
geography	[dʒɪˈɒgrəfɪ]	*n.* 地理（学）；地形，地势；布局
contiguous	[kənˈtɪgjʊəs]	*adj.* 接触的，邻近的；共同的
nonoverlapping	[ˈnɒnəʊvəˈlæpɪŋ]	*adj.* 不相重叠的
triangle	[ˈtraɪæŋgl]	*n.* 三角形
generalization	[ˌdʒenrəlaɪˈzeɪʃn]	*n.* 一般化，普通化；归纳，概论
conventional	[kənˈvenʃənl]	*adj.* 传统的；平常的；依照惯例的
shortcoming	[ˈʃɔːtkʌmɪŋ]	*n.* 短处，缺点
obstacle	[ˈɒbstəkl]	*n.* 障碍，障碍物
invariably	[ɪnˈveərɪəblɪ]	*adv.* 总是；不变的
attributable	[əˈtrɪbjʊtəbl]	*adj.* 可归因于……的；由……引起的
instantiate	[ɪnsˈtænʃɪeɪt]	*vt.* 例示
concretely	[ˈkɒŋkriːtlɪ]	*adv.* 具体地
interrelationship	[ˌɪntərɪˈleɪʃnʃɪp]	*n.* 相互关系，相互联系；影响，干扰
satisfy	[ˈsætɪsfaɪ]	*vt.* 符合，达到（要求、规定、标准等） *vi.* 使足够；使满意
resource	[rɪˈsɔːs]	*n.* 资源

Phrases

data model	数据模型
be composed of ...	由……组成
in turn	依次；轮流地；相应地；转而
a set of	一组，一套
data relationship	数据关系
data semantic	数据语义
data constraint	数据约束
programming language	编程语言，程序设计语言
function model	功能模型；函数模型
enterprise model	企业模型
information system	信息系统
data modeling language	数据建模语言
consist of ...	由……组成
object oriented	面向对象的
be concerned with ...	涉及……；与……相关
be consistent with ...	与……一致

carry out	执行，进行
be translated into	被转换为
database model	数据库模型
flat model	平面模型
table model	表模型
hierarchical model	层次模型，分层模型
network model	网络模型
tree structure	树状结构
base on...	基于……
first-order predicate logic	一阶谓词逻辑
finite set	有限集
mathematical foundation	数学基础
object-relational model	对象关系模型
attribute free	无属性
star schema	星型模式
data warehouse	数据仓库
fact table	事实表
dimension table	维度表
entity-relationship model	实体-关系模型
differ from	不同于
semantic data model	语义数据模型
physical data model	物理数据模型
software engineering	软件工程
geographic data model	地理数据模型
geographic information system	地理信息系统
generic data model	通用数据模型
conceptual data model	概念数据模型

Abbreviations

ANSI（American National Standards Institute）	美国国家标准学会
XML（extensible Markup Language）	可扩展标记语言
CPU（Central Processing Unit）	中央处理器
ERD（Entity Relationship Diagram）	实体关系图
UML（Unified Modeling Language）	统一建模语言，标准建模语言
DSD（Data Structure Diagram）	数据结构图

ERM（Entity Relationship Model） 实体关系模型
TIN（Triangulated Irregular Network） 不规则三角网络
DBMS（Database Management System） 数据库管理系统

Exercises

【Ex.1】Answer the following questions according to the text.

1. What is a data model?
2. What is the main aim of data models?
3. What does conceptual data model describe? What does it consist of?
4. What does physical data model describe? What is it concerned with?
5. What is the significance of the physical data model according to ANSI?
6. What is a database model? What are its common models mentioned in the passage?
7. What is a Data Structure Diagram (DSD)?
8. What is the difference between DSDs and ER models?
9. What is an Entity Relationship Model (ERM) sometimes referred to? What could it be used to do?
10. What is a semantic data model ?What is it in software engineering?

【Ex.2】 Translate the following terms or phrases from English into Chinese and vice versa.

1. function model	1. ________
2. hierarchical model	2. ________
3. star schema	3. ________
4. array	4. ________
5. function model	5. ________
6. 树状结构	6. ________
7. *n.* 收集，采集	7. ________
8. *n.* 形式化	8. ________
9. *n.* 范例，样式	9. ________
10. *n.* 一般化，普通化；归纳，概论	10. ________

【Ex.3】Translate the following passages into Chinese.

Types of Networks

Below is a list of the most common types of computer networks in order of scale.

1. Local Area Network (LAN)

Local area network is a network covering a small geographic area, like a home, office, or building. Current LANs are most likely to be based on Ethernet technology. The defining characteristics of LANs, in contrast to WANs (wide area networks), include their much higher data transfer rates, smaller geographic range, and lack of a need for leased telecommunication lines.

2. Campus Area Network (CAN)

Campus area network is a network that connects two or more LANs but that is limited to a specific (possibly private) geographical area such as a college campus, industrial complex, or a military base. A CAN may be considered a type of MAN (metropolitan area network), but is generally limited to an area that is smaller than a typical MAN.

3. Metropolitan Area Network (MAN)

MAN is a network that connects two or more LANs or CANs together but does not extend beyond the boundaries of the immediate town, city, or metropolitan area. Multiple routers, switches & hubs are connected to create a MAN.

4. Wide Area Network (WAN)

WAN is a data communications network that covers a relatively broad geographic area and that often uses transmission facilities provided by common carriers, such as telephone companies. WAN technologies generally function at the lower three layers of the OSI reference model: The physical layer, the data link layer, and the network layer.

【Ex.4】Fill in the blanks with the words given below.

physical	purpose	analysis	result	abstract
model	structures	database	share	involves

Data Modeling

1. Data Modeling Process

In the context of business process integration, data modeling complements business process modeling, and ultimately results in database generation.

The process of designing a database ____1____ producing the previously described three types of schemas-conceptual, logical, and physical. The database design documented in these schemas are converted through a Data Definition Language, which can then be used to generate a database. A fully attributed data ____2____ contains detailed attributes (descriptions) for every entity within it. The term "database design" can describe many different parts of the design of an overall database system. Principally, and most correctly, it can be thought of as the logical design of the base data ____3____ used to store the data. In the relational model these are the tables and views. In an object database the entities and relationships map directly to object classes and named relationships. However, the term "database design" could also be used to apply to the overall process of designing, not just the base data structures, but also the forms and queries used as part of the overall ____4____ application within the Database Management System or DBMS.

In the process, system interfaces account for 25% to 70% of the development and support costs of current systems. The primary reason for this cost is that these systems do not ___5___ a common data model. If data models are developed on a system by system basis, then not only is the same analysis repeated in overlapping areas, but further ___6___ must be performed to create the interfaces between them. Most systems within an organization contain the same basic data, redeveloped for a specific ___7___. Therefore, an efficiently designed basic data model can minimize rework with minimal modifications for the purposes of different systems within the organization.

2. Modeling Methodologies

Data models represent information areas of interest. While there are many ways to create data models, according to Len Silverston only two modeling methodologies stand out, top-down and bottom-up.

(1) Bottom-up models or View Integration models are often the ___8___ of a reengineering effort. They usually start with existing data structures forms, fields on application screens, or reports. These models are usually physical, application-specific, and incomplete from an enterprise perspective. They may not promote data sharing, especially if they are built without reference to other parts of the organization.

(2) Top-down logical data models, on the other hand, are created in an ___9___ way by getting information from people who know the subject area. A system may not implement all the entities in a logical model, but the model serves as a reference point or template.

Sometimes models are created in a mixture of the two methods: By considering the data needs and structure of an application and by consistently referencing a subject-area model. Unfortunately, in many environments the distinction between a logical data model and a ___10___ data model is blurred. In addition, some CASE tools don't make a distinction between logical and physical data models.

Text B
Data Model (2)

扫码听课文

4. Data Model Topics

4.1 Data Architecture

Data architecture is the design of data for use in defining the target state and the subsequent planning needed to hit the target state. It is usually one of several architecture domains that form the pillars of an enterprise architecture or solution architecture.

A data architecture describes the data structures used by a business and/or its applications. There are descriptions of data in storage and data in motion; descriptions of data stores, data groups, and data items; and mappings of those data artifacts to data qualities, applications, locations etc.

Data architecture describes how data is processed, stored, and utilized in a given system. It provides criteria for data processing operations that make it possible to design data flows and also control the flow of data in the system.

4.2 Data Modeling

Data modeling in software engineering is the process of creating a data model by applying formal data model descriptions using data modeling techniques. It is a technique for defining business requirements for a database. It is sometimes called database modeling because a data model is eventually implemented in a database.

Figure 2-7 illustrates the way data models are developed and used today. A conceptual data model is developed based on the data requirements for the application that is being developed, perhaps in the context of an activity model. The data model normally consists of entity types, attributes, relationships, integrity rules, and the definitions of those objects. This is then used as the start point for interface or database design.

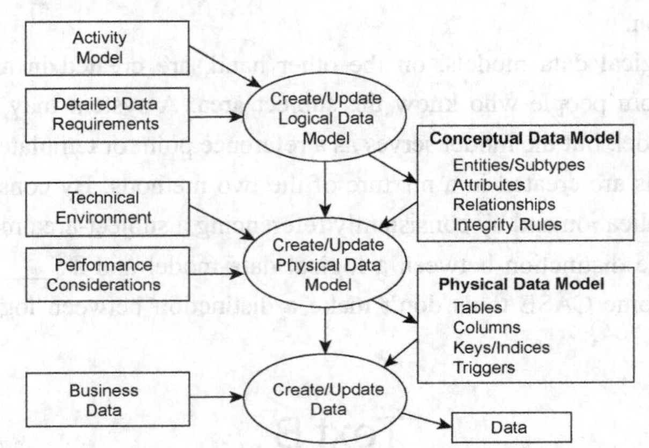

Figure 2-7 The Data Modeling Process

4.3 Data Properties

The following are some important properties of data (See Figure 2-8) for which requirements need to be met.

4.3.1 Definition-Related Properties

(1) Relevance: The usefulness of the data in the context of your business.

(2) Clarity: The availability of a clear and shared definition for the data.

(3) Consistency: The compatibility of the same type of data from different sources.

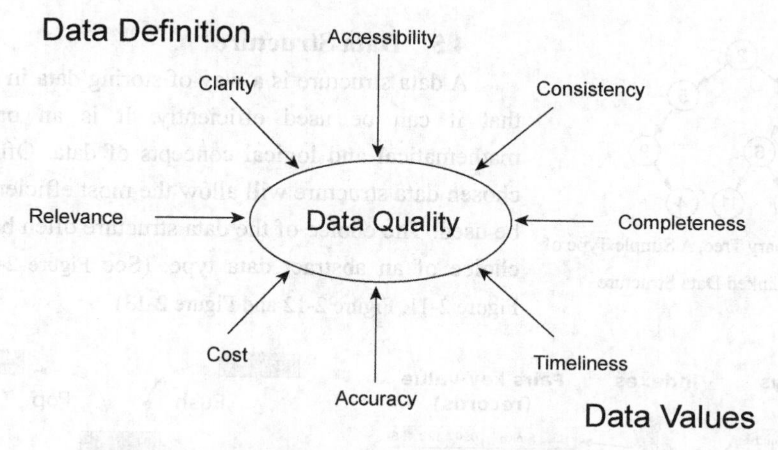

Figure 2-8 Some Important Properties of Data

4.3.2 Content-Related Properties

(1) Timeliness: The availability of data at the time required and how up to date that data is.
(2) Accuracy: How close to the truth the data is.
(3) Completeness: How much of the required data is available.
(4) Accessibility: Where, how, and to whom the data is available or not available (e.g. security).
(5) Cost: The cost incurred in obtaining the data, and making it available for use.

4.4 Data Organization

Another kind of data model describes how to organize data using a database management system or other data management technology. It describes, for example, relational tables and columns or object-oriented classes and attributes. Such a data model is sometimes referred to as the physical data model, but in the original ANSI three schema architecture, it is called "logical". In that architecture, the physical model describes the storage media (cylinders, tracks, and tablespaces). Ideally, this model is derived from the more conceptual data model described above. It may differ, however, to account for constraints like processing capacity and usage patterns.

While data analysis is a common term for data modeling, the activity actually has more in common with the ideas and methods of synthesis (inferring general concepts from particular instances) than it does with analysis (identifying component concepts from more general ones). Data modeling strives to bring the data structures of interest together into a cohesive, inseparable, whole by eliminating unnecessary data redundancies and by relating data structures with relationships.

A different approach is to use adaptive systems such as artificial neural networks that can autonomously create implicit models of data.

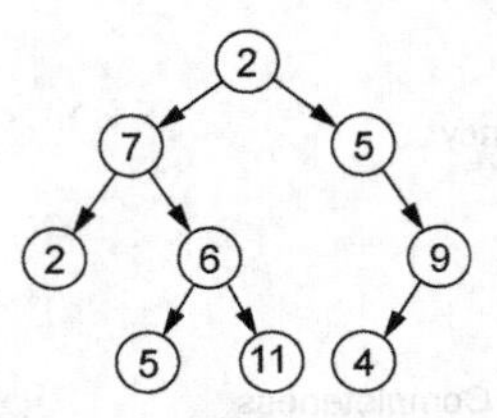

Figure 2-9 A Binary Tree, A Simple Type of Branching Linked Data Structure

4.5 Data Structure

A data structure is a way of storing data in a computer so that it can be used efficiently. It is an organization of mathematical and logical concepts of data. Often a carefully chosen data structure will allow the most efficient algorithm to be used. The choice of the data structure often begins from the choice of an abstract data type. (See Figure 2-9, Figure 2-10, Figure 2-11, Figure 2-12 and Figure 2-13)

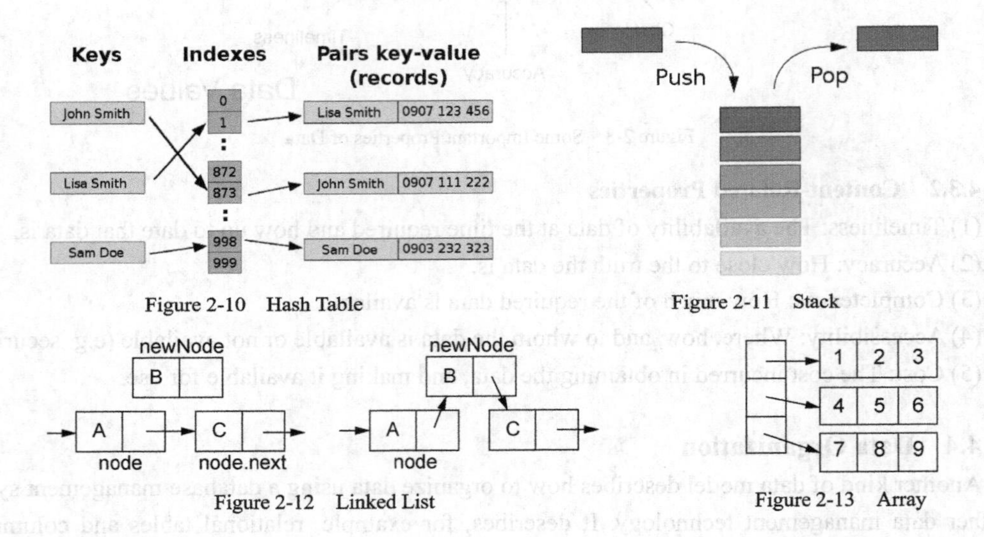

Figure 2-10 Hash Table

Figure 2-11 Stack

Figure 2-12 Linked List

Figure 2-13 Array

4.6 Data Model Theory

The term data model can have two meanings.

(1) A data model theory, i.e. , a formal description of how data may be structured and accessed.

(2) A data model instance, i.e., applying a data model theory to create a practical data model instance for some particular application.

A data model theory has three main components.

- The structural part: A collection of data structures which are used to create databases representing the entities or objects modeled by the database.
- The integrity part: A collection of rules governing the constraints placed on these data structures to ensure structural integrity.
- The manipulation part: A collection of operators which can be applied to the data structures, to update and query the data contained in the database.

For example, in the relational model, the structural part is based on a modified concept of the mathematical relation; the integrity part is expressed in first-order logic and the manipulation part is expressed using the relational algebra, tuple calculus, and domain calculus.

A data model instance is created by applying a data model theory. This is typically done to solve some business enterprise requirements. Business requirements are normally captured by a semantic logical data model. This is transformed into a physical data model instance from which a physical database is generated. For example, a data modeler may use a data modeling tool to create an entity-relationship model of the corporate data repository of some business enterprise. This model is transformed into a relational model, which in turn generates a relational database.

5. Related Models

5.1 Data Flow Diagram

A Data Flow Diagram (DFD) is a graphical representation of the "flow" of data through an information system. It differs from the flowchart as it shows the data flow instead of the control flow of the program. A data flow diagram can also be used for the visualization of data processing (structured design). Data flow diagrams were invented by Larry Constantine, the original developer of structured design, based on Martin and Estrin's "data flow graph" model of computation (See Figure 2-14).

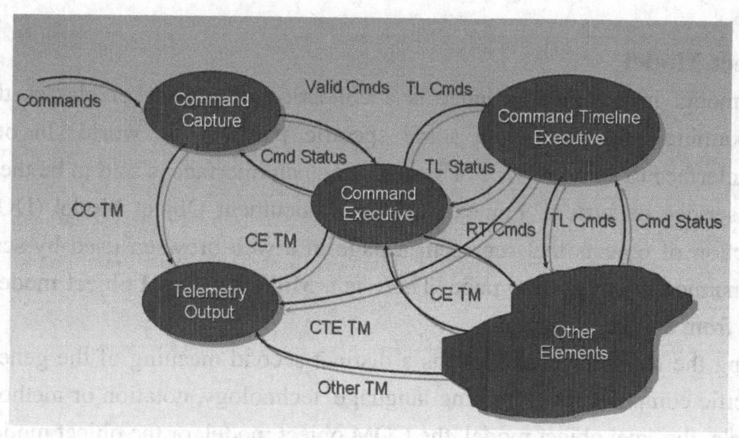

Figure 2-14 Data Flow Diagram Example

It is common practice to draw a context-level data flow diagram first which shows the interaction between the system and outside entitie. The DFD is designed to show how a system is divided into smaller portions and to highlight the flow of data between those parts. This context-level data flow diagram is then "exploded" to show more detail of the system being modeled.

5.2 Information Model

An information model is not a type of data model, but more or less an alternative model. Within the field of software engineering both a data model and an information model can be abstract, formal representations of entity types that include their properties, relationships, and the operations that can be performed on them. The entity types in the model may be kinds of real-world objects, such as devices in a network, or they may themselves be abstract, such as for the entities used in a billing system. Typically, they are used to model a constrained domain that can be described by a closed set of entity types, properties, relationships and operations.

According to Lee (1999) an information model is a representation of concepts, relationships, constraints, rules, and operations to specify data semantics for a chosen domain of discourse. It can provide sharable, stable, and organized structure of information requirements for the domain context. More in general the term information model is used for models of individual things, such as facilities, buildings, process plants, etc. In those cases the concept is specialised to facility information model, building information model, plant information model, etc. Such an information model is an integration of a model of the facility with the data and documents about the facility.

An information model provides formalism to the description of a problem domain without constraining how that description is mapped to an actual implementation in software. There may be many mappings of the information model. Such mappings are called data models, irrespective of whether they are object models, entity relationship models or XML schemas.

5.3 Object Model

An object model in computer science is a collection of objects or classes through which a program can examine and manipulate some specific parts of its world. In other words, the object-oriented interface to some service or system. Such an interface is said to be the object model of the represented service or system. For example, the Document Object Model (DOM) (See Figure 2-13) is a collection of objects that represent a page in a web browser, used by script programs to examine and dynamically change the page. There is a Microsoft Excel object model for controlling Microsoft Excel from another program.

In computing the term object model has a distinct second meaning of the general properties of objects in a specific computer programming language, technology, notation or methodology that uses them. For example, the Java object model, the COM object model, or the object model of OMT. Such object models are usually defined using concepts such as class, message, inheritance, polymorphism, and encapsulation. There is an extensive literature on formalized object models as a subset of the formal semantics of programming languages.

5.4 Object-Role Model

Object-Role Modeling (ORM) is a method for conceptual modeling, and can be used as a tool for information and rules analysis.

Object-role modeling is a fact-oriented method for performing systems analysis at the conceptual level. The quality of a database application depends critically on its design. To help ensure correctness, clarity, adaptability, and productivity, information systems are best specified first at the conceptual level, using concepts and language that people can readily understand.

The conceptual design may include data, process and behavioral perspectives, and the actual DBMS used to implement the design might be based on one of many logical data models (relational, hierarchic, network, object-oriented, etc.).

5.5 Unified Modeling Language Models

The Unified Modeling Language (UML) is a standardized general-purpose modeling language in the field of software engineering. It is a graphical language for visualizing, specifying, constructing, and documenting the artifacts of a software-intensive system. It offers a standard way to write a system's blueprints, including:

- Conceptual things such as business processes and system functions;
- Concrete things such as programming language statements, database schemas, and;
- Reusable software components.

UML also offers a mix of functional models, data models, and database models.

New Words

target	[ˈtɑːgɪt]	*n.* 目标，目的
subsequent	[ˈsʌbsɪkwənt]	*adj.* 后来的；随后的
pillar	[ˈpɪlə]	*n.* 台柱，顶梁柱；（组织、制度等的）核心
application	[ˌæplɪˈkeɪʃn]	*n.* 应用，适用
process	[ˈprəʊses]	*n.* 过程；工序；做事方法 *vt.* 加工；处理
criteria	[kraɪˈtɪərɪə]	*n.* 标准，准则
eventually	[ɪˈventʃʊəlɪ]	*adv.* 终于，最后；总归；终究
perhaps	[pəˈhæps]	*adv.* 或许，也许 *n.* 假定；猜想
interface	[ˈɪntəfeɪs]	*n.* 界面；接口
relevance	[ˈreləvəns]	*n.* 相关性，关联；实用性
usefulness	[ˈjuːsfʊlnɪs]	*n.* 有用，有益，有效
clarity	[ˈklærɪtɪ]	*n.* 清楚，明晰；透明；明确；清晰度
availability	[əˌveɪləˈbɪlɪtɪ]	*n.* 有效；有益；可利用性
compatibility	[kəmˌpætəˈbɪlɪtɪ]	*n.* 适合；互换性；通用性
timeliness	[ˈtaɪmlɪnɪs]	*n.* 及时性
accuracy	[ˈækjərəsi]	*n.* 精确（性），准确（性）

completeness	[kəm'pliːtnɪs]	*n.* 完全；完全性，完整性
cylinder	['sɪlɪndə]	*n.* 柱面
track	[træk]	*n.* 磁道
capacity	[kə'pæsətɪ]	*n.* 容量；能力 *adj.* 充其量的，最大限度的
synthesis	['sɪnθəsɪs]	*n.* 综合；综合体
infer	[ɪn'fɜː]	*vt.* 推断；猜想；暗示；意指 *vi.* 作出推论
strive	[straɪv]	*vi.* 努力奋斗，力求；力争
cohesive	[kəʊ'hiːsɪv]	*adj.* 紧密结合的
inseparable	[ɪn'seprəbl]	*adj.* 不可分的，分不开的；不能分离的
redundancy	[rɪ'dʌndənsɪ]	*n.* 冗余，过多，过剩
autonomously	[ɔː'tɒnəməslɪ]	*adv.* 自主地，独立地
implicit	[ɪm'plɪsɪt]	*adj.* 内含的；成为一部分的
express	[ɪk'spres]	*vt.* 表达
calculus	['kælkjʊləs]	*n.* 运算
flowchart	['fləʊtʃɑːt]	*n.* 流程图，作业图
visualization	[ˌvɪʒʊəlaɪ'zeɪʃn]	*n.* 可视化
interaction	[ˌɪntər'ækʃn]	*n.* 合作；互相影响；互动
alternative	[ɔːl'tɜːnətɪv]	*adj.* 替代的；备选的
device	[dɪ'vaɪs]	*n.* 装置，设备；方法，手段
discourse	['dɪskɔːs]	*n.* 论述，交谈；正式的讨论 *vi.* 谈论 *vt.* 叙述；讨论
sharable	['ʃeərəbl]	*adj.* 可分享的，可分担的
stable	['steɪbl]	*adj.* 稳定的
formalism	['fɔːməlɪzəm]	*n.* 形式主义 *adj.* 形式主义的
irrespective	[ˌɪrɪ'spektɪv]	*adj.* 不考虑的，不顾的；无关的
browser	['braʊzə]	*n.* 浏览器
dynamically	[daɪ'næmɪklɪ]	*adj.* 动态的
methodology	[ˌmeθə'dɒlədʒɪ]	*n.* 方法学，方法论
polymorphism	[ˌpɒlɪ'mɔːfɪzəm]	*n.* 多态性，多型现象
encapsulation	[ɪnˌkæpsjʊ'leɪʃn]	*n.* 封装
critically	['krɪtɪklɪ]	*adv.* 严重地
adaptability	[əˌdæptə'bɪlɪtɪ]	*n.* 适应性；合用性
readily	['redɪlɪ]	*adv.* 快捷地；轻而易举地；便利地

behavioral	[bɪˈheɪvɪərəl]	*adj.* 行为的，动作的
blueprint	[ˈbluːprɪnt]	*n.* 蓝图，设计图；计划大纲 *vt.* 为……制蓝图；为……制订计划
reusable	[ˌriːˈjuːzəbl]	*adj.* 可再用的，可重复使用的，可重用的

Phrases

data architecture	数据架构，数据体系结构
data flow	数据流
entity type	实体类型
integrity rule	完整性规则
data management technology	数据管理技术
artificial neural network	人工神经网络
abstract data type	抽象数据类型
a collection of	一组，一批；……的集合
be invented by ...	由……发明
be divided into	被分为
fact-oriented method	面向事实模型
modeling language	建模语言
software-intensive system	软件密集系统

Abbreviations

DFD（Data Flow Diagram）	数据流图
DOM（Document Object Model）	文档对象模型
COM（Component Object Model）	组件对象模型
OMT（Object Modeling Technique）	对象建模技术
ORM（Object-Role Modeling）	对象-角色建模

Exercises

【Ex.5】Answer the following questions according to the text.

1. What is data architecture?

2. What is data modeling? Why is it sometimes called database modeling?
3. What are the definition-related properties and content-related properties of data?
4. What does data modeling strive to do? How?
5. What is a data structure?
6. How many meanings can the term data model have? What are they? How many main components does a data model theory have? What are they?
7. What is Data Flow Diagram (DFD)? How does it differ from the flowchart?
8. What is an information model according to Lee (1999)? What can it provide?
9. What is an object model in computer science?
10. What is the Unified Modeling Language (UML) ?

参考译文

数据模型（1）

1. 概述

数据模型是一种抽象模型，它组织数据元素使之标准化，数据模型反映数据元素与现实世界实体的关系。例如，数据模型可以指定表示汽车的数据元素而其又由多种其他元素组成，这些元素代表汽车颜色和大小并定义其车主。

数据模型这一术语具有两种截然不同但密切相关的意义。有时它是指在特定应用领域中找到的对象和关系的抽象形式化，例如在制造业组织中找到的客户、产品和订单。在其他时候，它指的是用于定义这种形式化的一组概念：例如实体、属性、关系或表格等概念。因此，可以使用实体关系“数据模型”来定义银行应用程序的“数据模型”。本文在这两种意义上使用该术语。

数据模型基于数据、数据关系、数据语义和数据约束。数据模型提供要存储信息的细节。

数据模型有时可称为数据结构，尤其是在编程语言的情景中。数据模型通常由功能模型补充，特别是在业务模型的情景中。

数据模型的主要目的是通过提供数据的定义和格式支持信息系统的开发。

数据模型明确地确定数据的结构。数据模型的典型应用包括数据库模型、信息系统设计和数据交换。通常，数据模型由数据建模语言定义。

2. 数据模型的三个视角

1975 年，ANSI 确定数据模型实例可以是以下三种类型之一（见图 2-1）。

（1）概念数据模型：它描述了一个域的语义，即模型的范围。例如，它可以是组织或行业感兴趣领域的模型。它由实体类组成，表示域中重要的各类事物，以及各对实体类之间关联的关系断言。概念模式指定了可以使用模型表达的事实或命题的种类。从这个意义上讲，它定义了一个人工“语言”中允许的表达式，其范围受到模型范围的限制。

（2）逻辑数据模型：它描述了语义，由特定的数据操作技术表示。这包括表格和列的描述、面向对象的类和 XML 标记等。

（3）物理数据模型：它描述了存储数据的物理方法。这涉及分区、CPU、表格空间及类似的东西。

根据 ANSI 的说法，这种方法的重要性在于它允许三个视角相对独立。存储技术可以在不影响逻辑模型或概念模型的情况下进行更改。表/列结构可以在不（必然）影响概念模型的情况下进行更改。当然，在每种情况下，结构必须与其他模型保持一致。表/列结构可能不同于实体类和属性的直接转换，但它必须最终实现概念实体类结构的目标。许多软件开发项目的早期阶段都强调概念数据模型的设计。这种设计可以详细描述为逻辑数据模型。在后期，该模型可以转换为物理数据模型，但是，也可以直接实现概念模型。

3. 数据模型的类型

3.1 数据库模型

数据库模型是描述数据库的结构和使用方式的规范。已经提出了几种这类的模型。常见的模型如下。

3.1.1 平面模型

这可能不完全符合数据模型的要求。平面（或表）模型由数据元素的单个二维数组组成，其中假定给定列的所有成员都是相似的值，并假定给定行的所有成员彼此相关（见图 2-2）。

3.1.2 层次模型

层次模型（见图 2-3）类似于网络模型，只是层次模型中的链接形成树结构，而网络模型允许任意图形。

3.1.3 网络模型

该模型使用两个基本结构组织数据，称为记录和集合。记录包含字段，集合定义记录之间的一对多关系：一个所有者，多个成员。网络数据模型是数据库实现中使用的设计概念的抽象（见图 2-4）。

3.1.4 关系模型

关系模型是基于一阶谓词逻辑的数据库模型。它的核心思想是将数据库描述为有限的谓词变量集上的谓词集合，描述对可能值和值组合的约束。关系数据模型的强大之处在于其数学基础和简单的用户级别范例（见图 2-5）。

3.1.5 对象关系模型

与关系数据库模型类似，但在数据库模式和查询语言中直接支持对象、类和继承。

3.1.6 对象角色建模

这是一种数据建模方法，已被定义为“无属性”和“基于事实”。结果是一个可验证的正确系统，可以从中导出其他常见工件，例如 ERD、UML 和语义模型。在数据库设计过程期间描述数据对象之间的关联，使该过程的结果规范化。

3.1.7 星型模式

它是最简单的数据仓库模式。星型模式（见图 2-6）由几个引用任意数量的“维度表”的“事实表”组成。

3.2 数据结构图

数据结构图（DSD）是一种图表和数据模型，用于通过提供记录实体及其关系并绑定它们约束的图形符号来描述概念数据模型。DSD 的基本图形元素是表示实体的框和表示关系的箭头。数据结构图对于记录复杂数据实体最有用。

数据结构图是实体关系模型（ER 模型）的扩展。在 DSD 中，属性在实体框内而不是在之外指定，而关系则被绘制为由属性组成的框，这些属性指定将实体绑定在一起的约束条件。DSD 与 ER 模型的不同之处在于，ER 模型侧重于不同实体之间的关系，而 DSD 侧重于实体内元素的关系，使用户能够完全看到每个实体之间的链接和关系。

数据结构图的表示有多种样式，定义基数的方式有显著差异。选择包括箭头、倒箭头或表示基数的数字。

3.3 实体关系模型

实体关系模型（ERM），有时称为实体关系图（ERD），可用于表示抽象概念数据模型（或语义数据模型或物理数据模型），这在软件工程中用于表示结构化数据。有几种用于 ERM 的符号。与 DSD 一样，属性在实体框内而不是在之外指定，而关系则绘制为线条，关系约束作为线条上的描述。当表示具有多个属性的实体时，ER 模型虽然有力，但看起来可能很笨拙。

3.4 地理数据模型

地理信息系统中的数据模型是用于将地理对象或表面表示为数据的数学构造。例如：

（1）矢量数据模型将地理区域表示为点、线和多边形的集合；

（2）栅格数据模型将地理区域表示为存储数值的单元矩阵；

（3）并且，不规则三角网络（TIN）数据模型将地理位置表示为连续的非重叠三角形。

3.5 通用数据模型

通用数据模型是传统数据模型的概括。它们定义了标准化的一般关系类型，以及可能与这种类型相关联的事物。通用数据模型的开发解决了传统数据模型的一些缺点。例如，面对相同域，不同建模者通常使用不同的传统数据模型。这可能导致不同人员的模型难以结合在一起，并且造成数据交换和数据集成的障碍。然而，这种差异总是归因于模型中不同的抽象层次以及可以实例化的事实种类的差异。建模者需要沟通并就某些要更具体地呈现的元素达成一致，以使差异不那么显着。

3.6 语义数据模型

语义数据模型是一种抽象模型，它定义了存储的符号与现实世界的关系。语义数据模型有时被称为概念数据模型。软件工程中的语义数据模型是一种技术，它在与其他数据的相互关系的情景中定义数据含义。

数据库管理系统（DBMS）的逻辑数据结构，无论是分层的、网络的还是关系的，都不能完全满足数据概念定义的要求，因为它的范围有限并且偏向于 DBMS 采用的实现策略。因此，从概念上定义数据的需要导致了语义数据建模技术的发展。也就是说，它是在与其他数据的相互关系的情景中定义数据含义的技术。在资源、想法、事件等方面，现实世界在物理数据存储中被符号化地加以定义。语义数据模型是一种抽象，它定义了存储的符号与现实世界的关系。因此，模型必须是现实世界的真实表现。

Unit 3

Text A
Structured Data, Semi-Structured Data, and Unstructured Data

1. What Is Structured Data

Structured data has a high level of organization, which makes it predictable, easy to organize and very easily searchable using basic algorithms. The information is rigidly arranged. Data is entered in specific fields containing textual or numeric data. These fields often have their maximum or expected size defined. In addition to the firm structure for information, structured data has very set rules concerning how to access it.

Examples of structured data include relational databases and other transactional data like sales records, as well as Excel files that contain customer address lists. This type of data is generally stored in tables. You end up with various columns and rows of data. One column might be customer names, and other rows would contain further attributes such as address, zip code, phone, email, credit card number, etc.

2. What Is Unstructured Data

Unstructured data, on the other hand, is not organized in any discernable manner and has no associated data model. Some refer to data lakes as being the place where unstructured data is stored. This type of information is usually text-heavy and often includes multiple types of data. Examples of types of files generally considered to be unstructured data are: Books, some health records, satellite images, Adobe PDF files, a warranty request created by a customer service representative, notes in a web form, objects from presentations, blogs, text messages, word documents, videos, photos, and other images. These files are not organized other than being placed into a file system, object store or another repository.

3. What Is Semi-Structured Data

Matthew Magne, manager of Global Product Marketing for Data Management at SAS, defines semi-structured data as a type of data that contains semantic tags, but does not conform to the structure associated with typical relational databases. While semi-structured entities belong to the same class, they may have different attributes. Examples include email, XML and other markup languages.

While semi-structured data is not a natural fit for legacy databases, it is a critical source for big data analytics.

4. Where Does Semi-Structured Data Fit in

Semi-structured data falls in the middle between structured and unstructured data. It contains certain aspects that are structured, and others that are not. For example, X-rays and other large images consist largely of unstructured data — in this case, a great many pixels. It is impossible to search and query these X-rays in the same way that a large relational database can be searched, queried and analyzed. After all, all you are searching against are pixels within an image. Fortunately, there is a way around this. Although the files themselves may consist of no more than pixels, words or objects, most files include a small section known as metadata. This opens the door to being able to analyze unstructured data.

5. How Does Metadata Help

Metadata can be defined as a small portion of any file that contains data about the contents of the file. This often includes how the data was created, its purpose, its time of creation, the author, file size, length, sender/recipient, and more. As a result, large amounts of unstructured or semi-structured data can be catalogued, searched, queried and analyzed via their metadata.

X-rays and other image files also contain metadata. Queries against metadata could uncover the identity of the patient/doctor, the diagnosis, etc. Semi-structured data, then, is no longer useless to the business. On the contrary, it is now possible to mine great insight from it about customer habits, preferences, and opportunities.

6. How Does Unstructured and Semi-Structured Data Differ

If almost all unstructured data actually contains some kind of structure in the form of metadata, what's the difference? The reality is that there is a grey area between truly unstructured data and semi-structured data. Semi-structured data may lack organization and certainly is a million miles away from the rigorous organization of the information contained in a relational database. But the presence of metadata really makes the term semi-structured more appropriate than unstructured.

Very little data in the modern age has absolutely no structure and no metadata. In popular usage, therefore, most of what is termed unstructured data is really semi-structured data. Documents, images, and other files have some form of data structure. But for the sake of simplicity, data is loosely split into structured and unstructured categories. Some argue that the distinction between unstructured and semi-structured data is moot.

7. How Much is Semi-Structured Data out There

Unstructured and semi-structured data account for the vast majority of all data. Just consider the huge number of video files, audio files, and social media postings being added every minute, and you get an idea why the term big data originated.

Unstructured and semi-structured data represent 85% or more of all data. This percentage is only going to grow once machine learning, Artificial Intelligence (AI), and the Internet of Things (IoT) gain real momentum in the marketplace. That will lead to huge amounts of data flooding systems every second. For example, IoT sensors are expected to number tens of billions within the next five years. That's going to generate a lot of unstructured and semi-structured data.

8. How Does Big Data Fit in

It is not necessarily the size of the data that makes it big so much as the complexity of that data. Unstructured data is more complex and difficult to work with. Therefore, it is typically associated with big data. However, the reality is that big data contains a combination of structured, unstructured and semi-structured data. This combination further adds to the complexity.

Now big data technologies are such like Hadoop, NoSQL or MongoDB emerge. These relatively new technologies relax the usual data model requirements and allow the storing of data in a much more unstructured format.

But big data is only going to get bigger. Floods of semi-structured and unstructured data are already coming from the IoT, satellite imagery, digital microscopy, sonar explorations, Twitter feeds, Facebook, YouTube postings, and so on.

9. How Will Big Data Be Managed

Massive amounts of data are being created every second from a myriad of different file types. With millions of users demanding instant access, the management of big data becomes extremely challenging.

Whatever the storage mechanism, whether it is a data warehouse or a data lake, and however data is stored, big data entails a combination of structured and unstructured data. It all requires some level of data governance. Due to the sheer quantity of data involved, prioritization becomes vital, as well as alignment with business objectives.

10. Understanding Big Data

Big data can be best understood by considering four Vs: Volume, velocity, variety, and value.

Big data systems must be able to process the required volumes of data with sufficient velocity (both in terms of creation and distribution of that data). Further, systems must be able to cope with a wide variety of file types and data structures. With all of these elements in place, there is now an opportunity to extract real value from this information via analytics. The organizations that can manage all four Vs effectively stand to gain competitive advantage.

New Words

predictable	[prɪ'dɪktəbl]	*adj.* 可预测的，可预报的，可预见的
searchable	['sɜːtʃəbl]	*adj.* 可检索的
rigidly	['rɪdʒɪdlɪ]	*adv.* 严格地
arrange	[ə'reɪndʒ]	*vt.* 排列；把……（系统地）分类；整理
textual	['tekstʃuəl]	*adj.* 文本的
transactional	[træn'zækʃənəl]	*adj.* 交易的，业务的
discernable	[dɪ'sɜːnəbl]	*adj.* 可辨别的，可认识的
text-heavy	[tekst-'hevɪ]	*adj.* 文本为主的，文本化的
multiple	['mʌltɪpl]	*adj.* 多重的；多个的；多功能的
blog	[blɒg]	*n.* 博客
repository	[rɪ'pɒzɪtərɪ]	*n.* 仓库；贮藏室
conform	[kən'fɔːm]	*vi.* 符合；遵照；适应环境 *vt.* 使遵守；使一致 *adj.* 一致的
pixel	['pɪksl]	*n.* 像素
metadata	['metədeɪtə]	*n.* 元数据
sender	['sendə]	*n.* 寄件人，发送人
recipient	[rɪ'sɪpɪənt]	*n.* 收信人，接收者
uncover	[ʌn'kʌvə]	*vi.* 发现，揭示
portion	['pɔːʃn]	*n.* 一部分 *vt.* 分配
habit	['hæbɪt]	*n.* 习惯，习性
preference	['prefərəns]	*n.* 偏爱；优先权
rigorous	['rɪgərəs]	*adj.* 严密的；缜密的；严格的
absolutely	['æbsəluːtlɪ]	*adv.* 绝对地；完全地；毫无疑问地
simplicity	[sɪm'plɪsɪtɪ]	*n.* 简单，朴素

argue	[ˈɑːgjuː]	*vt.* 坚决主张；提出理由证明；说服，劝告；表明，证明
		vi. 争论；提出理由
distinction	[dɪˈstɪŋkʃn]	*n.* 区别
moot	[muːt]	*adj.* 无实际意义的
huge	[hjuːdʒ]	*adj.* 巨大的，庞大的，极大的
sensor	[ˈsensə]	*n.* 传感器
generate	[ˈdʒenəreɪt]	*vt.* 形成，造成；产生
complexity	[kəmˈpleksətɪ]	*n.* 复杂性
combination	[ˌkɒmbɪˈneɪʃn]	*n.* 组合，结合；联合体
manifest	[ˈmænɪfest]	*vt.* 显示，表明，显现
		adj. 明白的，明显的
sonar	[ˈsəʊnɑː]	*n.* 声呐装置，声呐系统
instant	[ˈɪnstənt]	*n.* 瞬间，顷刻
		adj. 立即的，即时的
challenge	[ˈtʃæləndʒ]	*n.* 挑战；质疑
		vt. 质疑；向……挑战
		vi. 提出挑战
prioritization	[praɪˌɒrətaɪˈzeɪʃn]	*n.* 优化；优先次序
vital	[ˈvaɪtl]	*adj.* 至关重要的；生死攸关的
sufficient	[səˈfɪʃnt]	*adj.* 足够的；充足的；充分的
distribution	[ˌdɪstrɪˈbjuːʃn]	*n.* 分配，分布
extract	[ˈekstrækt]	*vt.* 提取；选取；获得

Phrases

be entered in	被输入到
end up with...	以……结束
data lake	数据湖
web form	网页表单
semantic tag	语义标签，语义标记
markup language	标记语言，标识语言
fall in	落入，分成
be defined as	被定义为
grey area	灰色区域，模糊区域
split into	分成，分为

account for 占比重
vast majority 绝大多数
instant access 即时访问
competitive advantage 竞争优势

Abbreviations

SAS（Statistical Analysis System） 统计分析系统
IoT（Internet of Things） 物联网

Exercises

【Ex.1】Fill in the following blanks according to the text.

1. Structured data has a high level of organization, which makes it ________, easy to ________ and very easily ________using basic algorithms.
2. Examples of structured data include ________ and other transactional data like ________ as well as Excel files that contain ________. This type of data is generally stored in ________.
3. Unstructured data, on the other hand, is not organized ________ and has ________. Some refer to data lakes as being the place where ________ is stored.
4. Matthew Magne, manager of Global Product Marketing for Data Management at SAS, defines semi-structured data as a type of data that contains ________, but does not conform to the structure associated with ________.
5. Semi-structured data falls in the middle between ________ and ________ data. It contains certain aspects that are ________, and others that are not.
6. Metadata can be defined as a small portion of any file that contains ________. This often includes how the data was created, ________, ________, the author, ________, length, sender/recipient, and more.
7. Unstructured and semi-structured data represent ________ or more of all data. This percentage is only going to grow once ________, ________, and ________ gain real momentum in the marketplace.
8. It is not necessarily the size of the data that makes it big so much as ________ of that data. Unstructured data is ________ and ________ to work with.
9. Whatever the storage mechanism, whether it is ________ or ________, and ________, big data entails a combination of structured and unstructured data.
10. Big data can be best understood by considering four Vs: ________, ________, ________, and ________. The organizations that can manage all four Vs effectively stand to ________.

【Ex.2】Translate the following terms or phrases from English into Chinese and vice versa.

1. data lake 1. ______
2. searchable 2. ______
3. structured data 3. ______
4. be defined as 4. ______
5. textual 5. ______
6. *n.*复杂性 6. ______
7. *n.*分配，分布 7. ______
8. *vt.*提取；选取；获得 8. ______
9. *vt.*排列；把……分类；整理 9. ______
10. *vt.*访问，存取 10. ______

【Ex.3】Translate the following passages into Chinese.

Data Lake

A data lake is a storage repository that holds a vast amount of raw data in its native format until it is needed. While a hierarchical data warehouse stores data in files or folders, a data lake uses a flat architecture to store data. Each data element in a lake is assigned a unique identifier and tagged with a set of extended metadata tags. When a business question arises, the data lake can be queried for relevant data, and that smaller set of data can then be analyzed to help answer the question.

Data lakes and data warehouses are both used for storing big data, but each approach has its own uses. Typically, a data warehouse is a relational database housed on an enterprise mainframe server or in the cloud. The data stored in a warehouse is extracted from various online transaction processing (OLTP) applications to support business analytics queries and data marts for specific internal business groups, such as sales or inventory teams.

Data warehouses are useful when there is a massive amount of data from operational systems that needs to be readily available for analysis. Because the data in a lake is often uncurated and can originate from sources outside of the company's operational systems, lakes are not a good fit for the average business analytics user.

【Ex.4】Fill in the blanks with the words given below.

specific	operations	requires	addresses	based
logical	values	algorithms	retrieval	tables

Data Structure

In computer science, a data structure is a data organization, management, and storage format that enables efficient access and modification. More precisely, a data structure is a collection of data ___1___, the relationships among them, and the functions or operations that can be applied to the data.

Data structures serve as the basis for abstract data types (ADT). "The ADT defines the ___2___ form of the data type. The data structure implements the physical form of the data type."

Different types of data structures are suited to different kinds of applications, and some are highly specialized to ___3___ tasks. For example, relational databases commonly use B-tree indexes for data retrieval, while compiler implementations usually use hash ___4___ to look up identifiers.

Data structures provide a means to manage large amounts of data efficiently for uses such as large databases and internet indexing services. Usually, efficient data structures are key to designing efficient ___5___. Some formal design methods and programming languages emphasize data structures, rather than algorithms, as the key organizing factor in software design. Data structures can be used to organize the storage and ___6___ of information stored in both main memory and secondary memory.

Data structures are generally ___7___ on the ability of a computer to fetch and store data at any place in its memory, specified by a pointer—a bit string, representing a memory address, that can be itself stored in memory and manipulated by the program. Thus, the array and record data structures are based on computing the ___8___ of data items with arithmetic operations, while the linked data structures are based on storing addresses of data items within the structure itself. Many data structures use both principles, sometimes combined in non-trivial ways (as in XOR linking).

The implementation of a data structure usually ___9___ writing a set of procedures that create and manipulate instances of that structure. The efficiency of a data structure cannot be analyzed separately from those operations. This observation motivates the theoretical concept of an abstract data type, a data structure that is defined indirectly by the ___10___ that may be performed on it, and the mathematical properties of those operations (including their space and time cost).

Text B
Big Data Storage

扫码听课文

Enterprises have a lot to consider when they map out a plan for big data storage. Let's look at some of these factors in more detail.

1. Sizing up Big Data Storage Demand

Once you've created your quarterly requirement for big data storage, look at ways to reduce it. Much of the data is junk after a day or two. Some is sacred, so it should be stored and encrypted, with a backup and archive.

Look at the spikiness of demand. The public cloud is ideal for storing short-life data, especially if it is bursty. Storage buckets can be created and deleted cheaply, and scale definitely isn't an issue.

Finally, big data sometimes isn't that big! I've worked with 100 petabyte farms. Yep, that's big! For someone using 10 TB of structured data, 100 TB seems large, but it will fit easily in the minimum Ceph cluster. Don't overstate your problem! Today, solutions for 100 TB are straightforward.

2. The Role of Object Storage

Big data is often conflated with object storage because object storage can handle odd object sizes easily, and it provides metadata structures that allow tremendous control of data. This is all true. Moreover, object storage is much cheaper than traditional RAID arrays. In fact, the most common object storage uses open source software and COTS hardware. Unbundled licensed software is also available economically.

Object storage appliances come with six to twelve drives, a server board, and fast networks, and increasingly, the networking will be RDMA-based 100 GbE or 200 GbE. Even so, drives are getting so fast that these network rates may still struggle to keep up with. We are on the edge of NVMe over Ethernet connectivity for object storage, which will bring a leap forward in latency and throughput.

There are also open source global file systems that have been used in financial systems and high-performance computing for years. They handle the scale needed, but don't have extended metadata and other flexible extensions.

3. Life Cycle Management

Getting data in and out of your big data storage pool is a much bigger challenge than setting up the pool itself. Building end-of-life tagging into your storage software is one way to manage it: A policy sets the destruct tag value at data object creation time. Figuring out the policy takes time, though, and it gets more complex when disposition options are increased to include moving the data to very cheap archiving tiers in the cloud.

The data flow model for big data, especially IoT-generated big data, is often portrayed in storage marketing infographics as "a great river with many tributaries coming together". From the storage farm perspective, however, all that joining together doesn't really happen. Data has to be broken down into usable chunks and stored appropriately. Sensor data, the typical content generated by IoT, might be broken into time-stamped chunks, making later disposal easy, while structured database entries may be stored directly into the master database, which has its own tools for tiering cold data.

To complicate this, we know that some big data is much more active than others. This active data probably needs to be guided, by policy, to faster storage such as NVMe SSDs.

4. Data Privacy Laws

GDPR is close upon us. You might be forgiven for thinking this is nothing to do with the US or

Asia, but the rules for handling EU personal data carry draconian penalties of 4% of global revenue per violation, and apply worldwide. So if I sell a bottle of Napa wine in the US to someone in France, then let his or her personal data leak, I'd be in real trouble!

GDPR is, in the end, common sense for handling critical and personal data. Everyone should be encrypting data at rest properly and so on. The rules cover governance, life cycle management, access and use as well as encryption.

You might heave a sigh of relief on learning your storage vendor is GDRP compliant, but the rules involve a major paradigm shift for the data owner (you!) as well as any data storers. If you haven't gone through a realignment process, you're not compliant!

A common misconception is that vendor-provided encryption solves your compliance requirements. Drive-based encryption, whether provided by a storage vendor or a cloud service provider, is not adequate for any of the data standards such as HIPAA, SOX, or GDPR. You as data owner must own the keys. Fortunately, there is encryption support in the cloud, but a better alternative altogether is to build it into workflows back on your servers or virtual machines.

5. Solid-State Drives

SSDs are changing all the rules in storage systems. From acting as caches between DRAM and persistent storage to bulk storage devices, SSDs improve storage performance by factors of around 1000x in random IO and 10x to 100x in bandwidth. This is essential with huge volumes of data, especially when using parallel processing such as Hadoop, or GPU acceleration.

With 100 TB SSDs just over the horizon, and all this performance, a few small storage appliances can work wonders. The minimum Ceph object store arrays are four nodes and even using a standard 1U server format could hold 1.2 PB of raw SSD capacity today. It would not be cheap, but it would be economical when performance is calculated in. Vendors have already announced plans for 1U petabyte appliances, including one from Intel using 32 Ruler drives — long, but narrow SSDs.

The rapid development in this space is why you shouldn't invest too heavily in the short term. Price points and all other metrics are changing over the next two years. Ensure that any future buys of appliances and drives will fit the cluster, so that otherwise useful gear isn't scrapped.

6. Compression

Suppose I offer to turn that 1 PB appliance into 5 PB. That's the average benefit you'll get from using compression software. SSDs have so much bandwidth that using some of it for compression of data written to an appliance in the background makes sense. Still, I'm strongly in favor of compression at data creation. This reduces network traffic throughout the data flow, saves on storage space, and reduces time-to-transmit by, you guessed it, 5X too! Source compression needs hardware support and that's just beginning to appear in the market.

"Rehydrating" data is a trivial process that uses few resources, so increasing storage capacity with compression quickly translates to savings. All-flash arrays usually include compression. The technology is also offered as software for appliances.

7. The Cloud Alternative

After all talk about hardware, letting cloud providers do all the work might be an attractive option. In fact, the big three cloud service providers — Amazon, Google, and Microsoft — all lead when it comes to implementing new architectures and software orchestration. The cloud is economical and geared to paying for just the level of scale you need at any point. Cloud services can handle storage load spikes, which are common in some data classes such as retail sensor data, for example. This reduces or at least delays in-house gear purchases of storage gear.

Getting performance levels comparable to in-house operations, though, is a challenge. Not all instances with the same CPU and memory combinations are equal. A highly tuned in-house cluster might even do much better.

Today, storage doesn't stop with actually writing data to a drive. We are seeing value-added data storage services such as encryption and compression, indexing, tag servicing, and other features. The giant cloud providers, especially AWS, are even building database structures such as the Hadoop file system into the toolkit. This allows them to "invisibly" deploy gear such as key/data storage drives similar to new Seagate and Huawei units to accelerate specific data structures.

New Words

junk	[dʒʌŋk]	*n.* 废旧物品 *vt.* 丢弃，废弃
encrypt	[ɪn'krɪpt]	*v.* 加密，将……译成密码
backup	['bækʌp]	*n.* 备份，备份文件 *adj.* 备份的，备用的
archive	['ɑːkaɪv]	*v.* 存档 *n.* 档案文件
spikiness	['spaɪkɪnɪs]	*n.* 易于触怒；尖刻；带有尖刺
bursty	['bɜːstɪ]	*adj.* 阵发性的，间歇的
bucket	['bʌkɪt]	*n.* 一桶（的量）；大量 *v.* 用桶装，用桶运
cluster	['klʌstə]	*n.* 丛；簇；群 *vi.* 丛生；群聚 *vt.* 使密集，使聚集
overstate	[ˌəʊvə'steɪt]	*vt.* 夸大（某事）；把……讲得过分；夸张

straightforward	[ˌstreɪt'fɔːwəd]	*adj.* 简单的，易懂的
tremendous	[trə'mendəs]	*adj.* 极大的，巨大的；惊人的；极好的
economically	[ˌiːkə'nɒmɪklɪ]	*adv.* 节约地，节省地
struggle	['strʌgl]	*vi.* 努力；争取
		n. 竞争；奋斗
Ethernet	['iːθənet]	*n.* 以太网
connectivity	[ˌkɒnek'tɪvɪtɪ]	*n.* 连通性
throughput	['θruːpʊt]	*n.* 吞吐量；流率
destruct	[dɪs'trʌkt]	*adj.* 破坏的
		n. 故意或有计划的破坏
		vi. 破坏
disposition	[ˌdɪspə'zɪʃn]	*n.* 安排，配置
portray	[pɔː'treɪ]	*vt.* 描述，描绘
infographics	['ɪnfəʊ'græfɪks]	*n.* 信息图像
tributary	['trɪbjʊtərɪ]	*n.* 支流
appropriately	[ə'prəʊprɪətlɪ]	*adv.* 适当地
time-stamped	['taɪmstæmpt]	*n.* 时间戳，时间标记
violation	[ˌvaɪə'leɪʃn]	*n.* 违反，妨碍，侵犯；违犯，违背
realignment	[ˌriːə'laɪnmənt]	*n.* 改组
compliant	[kəm'plaɪənt]	*adj.* 遵从的；依从的；（与系列规则相）符合的；一致的
misconception	[ˌmɪskən'sepʃn]	*n.* 误解；错觉；错误想法
adequate	['ædɪkwɪt]	*adj.* 足够的；适当的
workflow	['wɜːkfləʊ]	*n.* 工作流程
cache	[kæʃ]	*n.* 快速缓冲存储区
persistent	[pə'sɪstənt]	*adj.* 持续的，持久的
random	['rændəm]	*adj.* 随机的
bandwidth	['bændwɪdθ]	*n.* 带宽
acceleration	[əkˌselə'reɪʃn]	*n.* 加速
node	[nəʊd]	*n.*（计算机网络的）节点
compression	[kəm'preʃn]	*n.* 压缩
rehydrate	[riː'haɪdreɪt]	*v.* 再水化，再水合
attractive	[ə'træktɪv]	*adj.* 有魅力的；引人注目的；迷人的；招人喜爱的
orchestration	[ˌɔːkɪ'streɪʃn]	*n.* 管弦乐编曲，管弦乐作曲法
delay	[dɪ'leɪ]	*n.* 延迟
toolkit	['tuːlkɪt]	*n.* 工具包，工具箱
invisibly	[ɪn'vɪzəblɪ]	*adv.* 看不见地，无形地
accelerate	[ək'seləreɪt]	*v.* 增速，加速

Phrases

big data storage	大数据存储
public cloud	公共云
short-life data	短期数据
conflated with ...	与……混杂在一起，与……混合在一起
unbundled licensed software	非捆绑式许可软件
network rate	网络速率，网速
keep up	持续不变；保持
life cycle	生命周期
storage pool	存储池
figuring out	搞清楚
storage farm	存储场
virtual machine	虚拟机
storage load spike	存储负载峰值

Abbreviations

TB（Terabyte）	太字节，1 TB=1024 GB
RAID（Redundant Arrays of Independent Drives）	磁盘阵列
COTS（Commercial Off-the-Shelf）	商用现成品或技术，商用货架产品
RDMA（Remote Direct Memory Access）	远程直接数据存取
GbE（Gigabit Ethernet）	千兆位以太网
NVMe（Non-Volatile Memory Express）	非易失性内存主机控制器接口规范
SSD（Solid-State Disk）	固态（磁）盘
GDPR（General Data Protection Regulation）	通用数据保护条例
HIPAA(Health Insurance Portability and Accountability Act）	健康保险携带和责任法案
SOX（Sarbanes-Oxley Act）	萨班斯法案
DRAM（Dynamic Random Access Memory）	动态随机存取存储器
GPU（Graphics Processing Unit）	图形处理器

Exercises

【Ex.5】Fill in the following blanks according to the text.

1. Once you've created your quarterly requirement for big data storage, look at ways to ________. Much of the data is junk ________. The public cloud is ideal for storing ________, especially if it

is bursty.

2. Big data is often ______ object storage because object storage can ______, and it provides ______ that allow tremendous control of data.

3. Object storage appliances come with ______, ______, and ______, and increasingly, the networking will be RDMA-based 100 GbE or 200 GbE.

4. The data flow model for big data, especially ______, is often portrayed in storage marketing infographics as ______.

5. GDPR is, in the end, common sense for handling ______. Everyone should be encrypting data at rest properly and so on. The rules cover ______, ______, ______ as well as encryption.

6. A common misconception is that vendor-provided encryption solves ______. Drive-based encryption, whether provided by ______ or ______, is not adequate for any of ______ such as HIPAA, SOX, or GDPR.

7. SSDs improve storage performance by factors of ______ and ______. This is essential with huge volumes of data, especially when using parallel processing such as ______, or ______.

8. "Rehydrating" data is a trivial process that uses ______, so increasing storage capacity with compression quickly translates to ______. All-flash arrays usually include ______. The technology also is offered as ______.

9. The cloud is ______ and geared to paying for just the level of scale you need at any point. Cloud services can handle ______, which are common in some data classes such as ______, for example. This reduces or at least delays ______ of storage gear.

10. Today, ______ doesn't stop with actually writing data to a drive. We are seeing value-added data storage services such as ______ and ______, ______, ______, and other features.

参考译文

结构化数据、半结构化数据和非结构化数据

1. 什么是结构化数据

结构化数据具有高度的组织性，使其可预测，易于组织，并且使用基本算法非常容易搜索。信息排列整齐。数据可输入到包含文本或数字数据的特定字段。这些字段通常定义了其最大或预期范围。除了固定的信息结构之外，结构化数据有设定的访问规则。

结构化数据的示例包括关系数据库和其他交易数据（如销售记录）以及包含客户地址列表

的 Excel 文件。这种类型的数据通常存储在表格中。最终得到各种以行、列显示的数据。一列可能是客户名称，其他行将包含更详细的属性，例如地址、邮政编码、电话、电子邮件、信用卡号等。

2. 什么是非结构化数据

非结构化数据没有以任何可识别的方式加以组织，也没有相关的数据模型。有些人将数据湖称为存储非结构化数据的地方。这类信息通常是以文本为主型的，同时往往包括多种类型的数据。通常被认为是非结构化数据文件类型的示例是：书籍、一些健康记录、卫星图像、Adobe PDF 文件、由客户服务代表创建的担保请求、网页表单中的注释、来自演示的对象、博客、文本消息、Word 文档、视频、照片和其他图像。除了放入文件系统、对象存储库或其他存储库之外，这些文件未经组织。

3. 什么是半结构化数据

SAS 数据管理全球产品营销部经理 Matthew Magne 给出的定义是：半结构化数据是一种包含语义标签的数据，但不符合与典型关系数据库相关的结构。虽然它与半结构化实体属于同一个类，但它们可能具有不同的属性。示例包括电子邮件、XML 和其他标记语言。

虽然半结构化数据不适合老式数据库，但它是大数据分析的关键来源。

4. 半结构化数据适合哪些方面

半结构化数据介于结构化和非结构化数据之间。它的某些方面是结构化的，而其他方面则不是。例如，X 射线图像和其他大型图像主要由非结构化数据组成——在这种情况下，有很多像素。要用搜索、查询和分析大型关系数据库的方法来搜索和查询这些 X 射线图像是不可能的，毕竟要搜索的只是图像中的像素。幸运的是，有一种解决方法，虽然文件本身可能只包含像素、单词或对象，但大多数文件都包含称为元数据的一小部分。这为分析非结构化数据打开了大门。

5. 元数据有何帮助

元数据可以定义为任何文件的一小部分，它包含了该文件内容的数据。这通常包括数据的创建方式、目的、创建时间、作者、文件大小、长度、发件人/收件人等。因此，可以通过元数据对大量非结构化或半结构化数据进行编目、搜索、查询和分析。

X 射线图像和其他图像文件也包含元数据，对元数据的查询可以揭示患者/医生的身份、诊断等。因此，半结构化数据对业务不再是无用的，相反，现在可以从中挖掘出有关客户习惯、偏好和机会的洞察性意见。

6. 非结构化和半结构化数据有何不同

如果几乎所有非结构化数据实际上都包含元数据形式的某种结构，那么二者有什么区别呢？实际情况是，真正的非结构化数据和半结构化数据之间存在灰色区域。半结构化数据可能缺乏组织，当然距离关系数据库中包含的信息的严格组织还差十万八千里。但元数据的存在确实使得半结构化术语比非结构化术语更合适。

现在，完全没有结构、也没有元数据的数据非常少。因此，在流行的用法中，大多数被称为非结构化数据的实际上是半结构化数据。文档、图像和其他文件具有某种形式的数据结构。但为了简便起见，数据被大致分为结构化和非结构化的。一些人认为区别非结构化和半结构化数据没有实际意义。

7．半结构化数据有多少

非结构化和半结构化数据占绝大多数。仅仅考虑每分钟添加的大量视频文件、音频文件和社交媒体帖子，就会明白大数据这个术语产生的原因。

非结构化和半结构化数据占所有数据的85%或更多。只要机器学习、人工智能（AI）和物联网（IoT）在市场上不断增长，这个百分比就会增长。这将导致每秒有大量的数据冲击系统。例如，物联网传感器预计在未来五年内将达到数百亿只，这将产生大量非结构化和半结构化数据。

8．如何适应大数据

与其说是数据的大小不如说是数据的复杂性使数据变大。非结构化数据更复杂、更难以使用。因此，它通常与大数据相关联。然而实际情况是，大数据包含了结构化、非结构化和半结构化数据的组合，这种组合进一步增加了复杂性。

现在出现了新兴的大数据技术，如Hadoop、NoSQL或MongoDB。这些相对较新的技术放宽了通常的数据模型要求，并允许以更非结构化的格式存储数据。

但大数据只会越来越大。物联网、卫星图像、数字显微镜、声纳探索、Twitter提要、Facebook YouTube帖子等已经致使半结构化和非结构化数据泛滥。

9．如何管理大数据

每秒都会从无数种不同的文件类型中产生出大量数据。由于数百万用户要求即时访问，大数据的管理变得极具挑战性。

无论存储机制是什么，无论是数据仓库还是数据湖以及无论数据是如何存储的，大数据都需要结构化和非结构化数据的组合。这一切都需要一定程度的数据治理。由于涉及的数据量很大，划分优先级变得至关重要，并且应与业务目标保持一致。

10．理解大数据

通过考虑4V可以更好地理解大数据：数量、高速、多样和价值。大数据系统必须能够以足够快的速度处理所需的数据量（在数据的创建和分发方面）。此外，系统必须能够处理各种各样的文件类型和数据结构。所有这些到位之后，现在就有机会通过进行分析从这些信息中提取真正的价值。能够有效管理所有4V的组织有望获得竞争优势。

Unit 4

Text A

ETL

ETL is a type of data integration that refers to the three steps (extract, transform, load) used to blend data from multiple sources. It's often used to build a data warehouse. During this process, data is taken (extracted) from a source system, converted (transformed) into a format that can be analyzed, and stored (loaded) into a data warehouse or other system. Extract, load, transform (ELT) is an alternate but related approach designed to push processing down to the database for improved performance.

1. Why Is ETL Important

For many years, businesses have relied on the ETL process to get a consolidated view of the data that drives better business decisions. Today, this method of integrating data from multiple systems and sources is still a core component of an organization's data integration toolbox.

- When used with an enterprise data warehouse (data at rest), ETL provides deep historical context for the business.
- By providing a consolidated view, ETL makes it easier for business users to analyze and report on data relevant to their initiatives.
- ETL can improve data professionals' productivity because it codifies and reuses processes that move data without requiring technical skills to write code or scripts.
- ETL has evolved over time to support emerging integration requirements for things like streaming data.
- Organizations need both ETL and ELT to bring data together, maintain accuracy and provide the auditing typically required for data warehousing, reporting, and analytics.

2. How Is ETL Being Used

Core ETL and ELT tools work in tandem with other data integration tools, and with various other aspects of data management — such as data quality, data governance, virtualization, and metadata. Popular uses today include the following.

2.1 ETL and Traditional Uses

ETL is a proven method that many organizations rely on every day — such as retailers who need to see sales data regularly, or health care providers looking for an accurate depiction of claims. ETL can combine and surface transaction data from a warehouse or other data store so that it's ready for business people to view in a format they can understand. ETL is also used to migrate data from legacy systems to modern systems with different data formats. It's often used to consolidate data from business mergers, and to collect and join data from external suppliers or partners.

2.2 ETL with Big Data — Transformations and Adapters

Whoever gets the most data, wins. While that's not necessarily true, having easy access to a broad scope of data can give businesses a competitive edge. Today, businesses need access to all sorts of big data — from videos, social media, the Internet of Things (IoT), server logs, spatial data, open or crowdsourced data, and more. ETL vendors frequently add new transformations to their tools to support these emerging requirements and new data sources. Adapters give access to a huge variety of data sources, and data integration tools interact with these adapters to extract and load data efficiently.

2.3 ETL for Hadoop — and More

ETL has evolved to support integration across much more than traditional data warehouses. Advanced ETL tools can load and convert structured and unstructured data into Hadoop. These tools read and write multiple files in parallel from and to Hadoop, simplifying how data is merged into a common transformation process. Some solutions incorporate libraries of prebuilt ETL transformations for both the transaction and interaction data that run on Hadoop. ETL also supports integration across transactional systems, operational data stores, BI platforms, master data management (MDM) hubs, and the cloud.

2.4 ETL and Self — Service Data Access

Self-service data preparation is a fast-growing trend that puts the power of accessing, blending and transforming data into the hands of business users and other nontechnical data professionals. This approach increases organizational agility and frees IT from the burden of provisioning data in different formats for business users. Less time is spent on data preparation and more time is spent on generating insights. Consequently, both business and IT data professionals can improve productivity, and organizations can scale up their use of data to make better decisions.

2.5 ETL and Data Quality

ETL and other data integration software tools — used for data cleansing, profiling, and auditing — ensure that data is trustworthy. ETL tools integrate with data quality tools, and ETL vendors incorporate related tools within their solutions, such as those used for data mapping and data lineage.

2.6 ETL and Metadata

Metadata helps us understand the lineage of data (where it comes from) and its impact on other data assets in the organization. As data architectures become more complex, it's important to track how the different data elements in your organization are used and related. For example, if you add a Twitter account name to your customer database, you'll need to know what will be affected, such as ETL jobs, applications or reports.

3. How It Works

ETL is closely related to a number of other data integration functions, processes, and techniques.

3.1 SQL

Structured query language is the most common method of accessing and transforming data within a database.

3.2 Transformations, Business Rules and Adapters

After extracting data, ETL uses business rules to transform the data into new formats. The transformed data is then loaded into the target.

3.3 Data Mapping

Data mapping is part of the transformation process. Mapping provides detailed instructions to an application about how to get the data it needs to process. It also describes which source field maps to which destination field. For example, the third attribute from a data feed of website activity might be the user name, the fourth might be the time stamp of when that activity happened, and the fifth might be the product that the user clicked on. An application or ETL process using that data would have to map these same fields or attributes from the source system (i.e., the website activity data feed) into the format required by the destination system. If the destination system was a customer relationship management system, it might store the user name first and the time stamp fifth; it might not store the selected product at all. In this case, a transformation to format the date in the expected format (and in the right order) might happen between the time the data was read from the source and written to the target.

3.4 Scripts

ETL is a method of automating the scripts (set of instructions) that run behind the scenes to move and transform data. Before ETL, scripts were written individually in C or COBOL to transfer data between specific systems. This resulted in multiple databases running numerous scripts. Early ETL tools ran on mainframes as a batch process. ETL later migrated to UNIX and PC platforms. Organizations today still use both scripts and programmatic data movement methods.

3.5 ETL vs. ELT

In the beginning, there was ETL. Later, organizations added ELT, a complementary method. ELT extracts data from a source system, loads it into a destination system and then uses the processing power of the source system to conduct the transformations. This speeds data processing because it happens where the data lives.

3.6 Data Quality

Before data is integrated, a staging area is often created where data can be cleansed, data values can be standardized (NC and North Carolina, Mister and Mr., or Matt and Matthew), addresses can be verified and duplicates can be removed. Many solutions are still standalone, but data quality procedures can now be run as one of the transformations in the data integration process.

3.7 Scheduling and Processing

ETL tools and technologies can provide either batch scheduling or real-time capabilities. They can also process data at high volumes on the server, or they can push down processing to the database level. This approach of processing in a database as opposed to a specialized engine avoids data duplication and prevents the need to use extra capacity on the database platform.

3.8 Batch Processing

ETL usually refers to a batch process of moving huge volumes of data between two systems during what's called a "batch window". During this set period of time — say between noon and 1 p.m. — no actions can happen to either the source or target system as data is synchronized. Most banks do a nightly batch process to resolve transactions that occur throughout the day.

3.9 Web Services

Web services are an internet-based method of providing data or functionality to various applications in near-real time. This method simplifies data integration processes and can deliver more value from data, faster. For example, let's say a customer contacts your call center. You could create a web service that returns the complete customer profile with a subsecond response time simply by passing a phone number to a web service that extracts the data from multiple sources or an MDM hub. With richer knowledge of the customer, the customer service rep can make better decisions about how to interact with the customer.

3.10 Master Data Management (MDM)

MDM is the process of pulling data together to create a single view of the data across multiple sources. It includes both ETL and data integration capabilities to blend the data together and create a "golden record" or "best record".

3.11 Data Virtualization

Virtualization is an agile method of blending data together to create a virtual view of data without moving it. Data virtualization differs from ETL, because even though mapping and joining data still occurs, there is no need for a physical staging table to store the results. That's because the view is often stored in memory and cached to improve performance. Some data virtualization solutions, like SAS Federation Server, provide dynamic data masking, randomization, and hashing functions to protect sensitive data from specific roles or groups. SAS also provides on-demand data quality while the view is generated.

3.12 Event Stream Processing and ETL

When the speed of data increases to millions of events per second, event stream processing can be used to monitor streams of data, process the data streams and help make more timely decisions.

New Words

integration	[ˌɪntɪ'greɪʃn]	*n.* 集成，整合，一体化
transform	[træns'fɔːm]	*vt.* 转换，改变，变换
load	[ləʊd]	*n.* 负荷；装载；工作量
		vi. 加载；装载
blend	[blend]	*vt.* 混合；（使）调和；协调
		n. 混合；混合物
convert	[kən'vɜːt]	*vt.*（使）转变
		vi. 经过转变；被改变
analyze	['ænəlaɪz]	*vt.* 分析，分解，解释
alternate	[ɔːl'tɜːnət]	*adj.* 交替的，代替的
		n. 候补者；替换物
	['ɔːltəneɪt]	*vi.* 交替；轮流
		vt. 使交替；使轮流
database	['deɪtəbeɪs]	*n.* 数据库
consolidate	[kən'sɒlɪdeɪt]	*vt.* 把……合成一体，合并
		vi. 统一；合并；联合
toolbox	['tuːlbɒks]	*n.* 工具箱
initiative	[ɪ'nɪʃətɪv]	*n.* 主动性；主动精神

		adj. 自发的；创始的
code	[kəʊd]	*n.* 代码
		vt. 编码，加密
		vi. 为……编码
script	[skrɪpt]	*n.* 脚本
accuracy	[ˈækjʊərəsɪ]	*n.* 精确（性），准确（性）
regularly	[ˈregjʊləlɪ]	*adv.* 有规律地；整齐地；不断地；定期地
migrate	[maɪˈgreɪt]	*vi.* 迁移，移动
collect	[kəˈlekt]	*vt.* 收集
crowdsourced	[ˈkraʊdsɔːst]	*adj.* 众包的
merge	[mɜːdʒ]	*vt.&vi.*（使）混合；相融，融入
prebuilt	[priːˈbɪlt]	*v.* 预建，预置
hub	[hʌb]	*n.* 中心
preparation	[ˌprepəˈreɪʃn]	*n.* 准备，预备
nontechnical	[nɒnˈteknɪkl]	*adj.* 非技术性的
agility	[əˈdʒɪlɪtɪ]	*n.* 敏捷，灵活
burden	[ˈbɜːdn]	*n.* 负担，包袱；责任，义务
trustworthy	[ˈtrʌstwɜːðɪ]	*adj.* 值得信赖的，可靠的
lineage	[ˈlɪnɪɪdʒ]	*n.* 血统，世系
track	[træk]	*vt.* 跟踪，追踪
instruction	[ɪnˈstrʌkʃn]	*n.* 指令
website	[ˈwebsaɪt]	*n.* 网站
scene	[ˈsiːn]	*n.* 背景，场景
individually	[ˌɪndɪˈvɪdʒʊəlɪ]	*adv.* 分别地，各个地，各自地
platform	[ˈplætfɔːm]	*n.* 平台
programmatic	[ˌprəʊgrəˈmætɪk]	*adj.* 程序化的
complementary	[ˌkɒmplɪˈmenterɪ]	*adj.* 互补的，补充的，补足的
standalone	[ˈstændəˌləʊn]	*n.* 脱机
		adj. 单独的
procedure	[prəˈsiːdʒə]	*n.* 程序，过程
duplication	[ˌdjuːplɪˈkeɪʃn]	*n.* 重复；复制
synchronize	[ˈsɪŋkrənaɪz]	*vt.* 使同步；使同时
		vi. 同时发生；共同行动
functionality	[ˌfʌŋkʃəˈnælɪtɪ]	*n.* 功能；功能性
subsecond	[səbˈsekənd]	*n.* 亚秒
virtualization	[vɜːtʃʊəlaɪˈzeɪʃn]	*n.* 虚拟化
randomization	[ˌrændəmaɪˈzeɪʃn]	*n.* 随机化，随机选择

Phrases

business decision	业务决策，商业决策
streaming data	流数据
in tandem with ...	与……合作
data quality	数据质量
data governance	数据治理
competitive edge	竞争优势
server logs	服务器日志
spatial data	空间坐标数据
crowdsourced data	众包数据
interact with ...	与……相互配合，与……相互影响
spend...on	花……在……
scale up	按比例提高，按比例增加
data cleansing	数据清理
data mapping	数据映射
data lineage	数据沿袭
business rule	业务规则，商业规则
data feed	数据馈送
time stamp	时间戳
customer relationship management	客户关系管理
batch process	批处理
staging area	临时区域
batch window	批处理窗口
staging table	临时表
dynamic data masking	动态数据屏蔽
hashing function	散列函数；杂凑函数；哈希函数
sensitive data	敏感数据
event stream processing	事件流处理

Abbreviations

ELT（Extract, Load, Transform）	提取、加载、转换
BI（Business Intelligence）	商业智能
MDM（Master Data Management）	主数据管理
SQL（Structured Query Language）	结构化查询语言

Exercises

【Ex.1】Answer the following questions according to the text.

1. What is ETL?
2. What have businesses relied on the ETL process to do for many years?
3. Why can ETL improve data professionals' productivity?
4. What can advanced ETL tools do?
5. As data architectures become more complex, what is it important to do?
6. What is data mapping? What does it do?
7. What can ETL tools and technologies do?
8. What are web services? What do they do?
9. What is MIDM? What does it include?
10. Why does data virtualization differ from ETL?

【Ex.2】Translate the following terms or phrases from English into Chinese and vice versa.

1. data mapping	1.
2. data cleansing	2.
3. analyze	3.
4. hashing function	4.
5. accuracy	5.
6. *vt.*收集	6.
7. 时间戳	7.
8. *n.*指令	8.
9. *n.*程序，过程	9.
10. *v.*复制	10.

【Ex.3】Translate the following passages into Chinese.

Data Quality

Data quality is a perception or an assessment of data's fitness to serve its purpose in a given context. The quality of data is determined by factors such as accuracy, completeness, reliability, relevance and how up to date it is. As data has become more intricately linked with the operations of organizations, the emphasis on data quality has gained greater attention.

Aspects, or dimensions, important to data quality include: Accuracy, or correctness; completeness, which determines if data is missing or unusable; conformity, adherence to a standard format; consistency, lack of conflict with other data values.

As a first step toward data quality, organizations typically perform data asset inventories in which the relative value, uniqueness, and validity of data can undergo baseline studies.

Methodologies for such data quality projects include the Data Quality Assessment Framework (DQAF), which is created by the International Monetary Fund (IMF) to provide a common method for assessing data quality. The DQAF provides guidelines for measuring data dimensions.

【Ex.4】Fill in the blanks with the words given below.

software	system	restoring	recovered	program
records	loss	database	transactions	updates

Recovery Process

Since business organizations have become very dependent on transaction processing, a breakdown may disrupt the business' regular routine and stop its operation for a certain amount of time. In order to prevent data ____1____ and minimize disruptions there have to be well-designed backup and recovery procedures. The recovery process can rebuild the system when it goes down.

A TPS may fail for many reasons such as system failure, human errors, hardware failure, incorrect or invalid data, computer viruses, ____2____ application errors or natural or man-made disasters. As it's not possible to prevent all failures, a TPS must be able to detect and correct errors when they occur and cope with failures. A TPS will go through a recovery of the database which may involve the backup, journal, checkpoint, and recovery manager.

- Journal: A journal maintains an audit trail of transactions and database changes. Transaction logs and Database change logs are used, a transaction log ____3____ all the essential data for each transactions, including data values, time of transaction and terminal number. A database change log contains before and after copies of records that have been modified by transactions.
- Checkpoint: The purpose of checkpointing is to provide a snapshot of the data within the database. A checkpoint, in general, is any identifier or other reference that identifies the state of the database at a point in time. Modifications to database pages are performed in memory and are not necessarily written to disk after every update. Therefore, periodically, the database system must perform a checkpoint to write these ____4____ which are held in-memory to the storage disk. Writing these updates to storage disk creates a point in time in which the database system can apply changes contained in a transaction log during recovery after an unexpected shut down or crash of the database ____5____. If a checkpoint is interrupted and a recovery is required, then the database system must start recovery from a previous successful checkpoint. Checkpointing can be either transaction-consistent or non-transaction-consistent (called also fuzzy checkpointing). Transaction-consistent checkpointing produces a persistent database image that is sufficient to recover the ____6____ to the state that was externally perceived at the moment of starting the checkpointing. A non-transaction-consistent checkpointing results in a persistent database image that is insufficient to

perform a recovery of the database state. To perform the database recovery, additional information is needed, typically contained in transaction logs. Transaction-consistent checkpointing refers to a consistent database, which doesn't necessarily include all the latest committed transactions, but all modifications made by ___7___, that were committed at the time checkpoint creation was started, are fully present. A non-consistent transaction refers to a checkpoint which is not necessarily a consistent database, and can't be ___8___ to one without all log records generated for open transactions included in the checkpoint. Depending on the type of database management system implemented a checkpoint may incorporate indexes or storage pages (user data), indexes and storage pages. If no indexes are incorporated into the checkpoint, indexes must be created when the database is restored from the checkpoint image.

- Recovery Manager: A recovery manager is a ___9___ which restores the database to a correct condition which allows transaction processing to be restarted.

Depending on how the system failed, there can be two different recovery procedures used. Generally, the procedures involves ___10___ data that has been collected from a backup device and then running the transaction processing again. Two types of recovery are backward recovery and forward recovery.

- Backward recovery: Used to undo unwanted changes to the database. It reverses the changes made by transactions which have been aborted.
- Forward recovery: It starts with a backup copy of the database. The transaction will then reprocess according to the transaction journal that occurred between the time the backup was made and the present time.

Text B
Data Backup

扫码听课文

In information technology, a data backup, or the process of backing up, refers to the copying into an archive file of computer data so it may be used to restore the original after a data loss event.

Backups have two distinct purposes. The primary purpose is to recover data after its loss, be it by data deletion or corruption. Data loss can be a common experience of computer users; a 2008 survey found that 66% of the respondents had lost files on their home PC. The secondary purpose of backups is to recover data from an earlier time, according to a user-defined data retention policy, typically configured within a backup application for how long copies of data are required. Though backups represent a simple form of disaster recovery and should be part of any disaster recovery plan, backups by themselves should not be considered a complete disaster recovery plan. One reason for this is that not all backup systems are able to reconstitute a computer system or other complex configuration such as a computer cluster, active directory server, or database server by simply restoring data from a backup.

Since a backup system contains at least one copy of all data considered worth saving, the data storage requirements can be significant. Organizing this storage space and managing the backup process can be a complicated undertaking. A data repository model may be used to provide structure to the storage. Nowadays, there are many different types of data storage devices that are useful for making backups. There are also many different ways in which these devices can be arranged to provide geographic redundancy, data security, and portability.

Before data are sent to their storage locations, they are selected, extracted, and manipulated. Many different techniques have been developed to optimize the backup procedure. These include optimization for dealing with open files and live data sources as well as compression, encryption, and de-duplication, among others. Every backup scheme should include dry runs that validate the reliability of the data being backed up. It is important to recognize the limitations and human factors involved in any backup scheme.

1. Storage, the Base of a Backup System

1.1 Data Repository Models

Any backup strategy starts with a concept of a data repository. The backup data needs to be stored, and probably should be organized to a degree. The organisation could be as simple as a sheet of paper with a list of all backup media (CDs, etc.) and the dates they were produced. A more sophisticated setup could include a computerized index, catalog, or relational database. Different approaches have different advantages. Part of the model is the backup rotation scheme.

1.1.1 Unstructured

An unstructured repository may simply be a stack of tapes or CD-Rs or DVD-Rs with minimal information about what is backed up and when. This is the easiest to implement, but probably the least likely to achieve a high level of recoverability as it lacks automation.

1.1.2 Full Only / System Imaging

A repository of this type contains complete system images taken at one or more specific points in time. This technology is frequently used by computer technicians to record known good configurations. Imaging is generally more useful for deploying a standard configuration to many systems rather than as a tool for making ongoing backups of diverse systems.

1.1.3 Incremental

An incremental style repository aims to make it more feasible to store backups from more points in time by organizing the data into increments of change between points in time. This eliminates the need to store duplicate copies of unchanged data: With full backups a lot of the data will be unchanged from what has been backed up previously. Typically, a full backup (of all files) is made on one occasion (or at infrequent intervals) and serves as the reference point for an incremental backup set. After that, a number of *incremental* backups are made after successive time periods. Restoring the whole system to the date of the last incremental backup would require starting from the last full backup taken before the data loss, and then applying in turn each of the incremental backups since then. Additionally, some backup systems can reorganize the repository to synthesize full backups from a series of incrementals.

1.1.4 Differential

Each differential backup saves the data that has changed since the last full backup. It has the advantage that only a maximum of two data sets are needed to restore the data. One disadvantage, compared to the incremental backup method, is that as time from the last full backup (and thus the accumulated changes in data) increases, so does the time to perform the differential backup. Restoring an entire system would require starting from the most recent full backup and then applying just the last differential backup since the last full backup.

Note: Vendors have standardized on the meaning of the terms "incremental backup" and "differential backup". However, there have been cases where conflicting definitions of these terms have been used. The most relevant characteristic of an incremental backup is which reference point it uses to check for changes. By standard definition, a differential backup copies files that have been created or changed since the last full backup, regardless of whether any other differential backups have been made since then, whereas an incremental backup copies files that have been created or changed since the most recent backup of any type (full or incremental). Other variations of incremental backup include multi-level incrementals and incremental backups that compare parts of files instead of just the whole file.

1.1.5 Reverse Delta

A reverse delta type repository stores a recent "mirror" of the source data and a series of differences between the mirror in its current state and its previous states. A reverse delta backup will start with a normal full backup. After the full backup is performed, the system will periodically synchronize the full backup with the live copy, while storing the data necessary to reconstruct older versions. This can either be done using hard links, or using binary diffs. This system works particularly well for large, slowly changing data sets.

1.1.6 Continuous Data Protection

Instead of scheduling periodic backups, the system immediately logs every change in the host system. This is generally done by saving byte or block-level differences rather than file-level differences. It differs from simple disk mirroring in that it enables a roll-back of the log and thus restoration of old images of data.

1.2 Managing the Data Repository

Regardless of the data repository model, or data storage media used for backups, a balance needs to be struck between accessibility, security, and cost. These media management methods are not mutually exclusive and are frequently combined to meet the user's needs. Using on-line disks for staging data before it is sent to a near-line tape library is a common example.

Data repository implementations include:

1.2.1 On-Line

On-line backup storage is typically the most accessible type of data storage, which can begin restore in milliseconds of time. A good example is an internal hard disk or a disk array (maybe connected to SAN). This type of storage is very convenient and speedy, but is relatively expensive. On-line storage is quite vulnerable to being deleted or overwritten, either by accident, by intentional malevolent action, or in the wake of a data-deleting virus payload.

1.2.2 Near-Line

Near-line storage is typically less accessible and less expensive than on-line storage, but still useful for backup data storage. A good example would be a tape library with restore times ranging from seconds to a few minutes. A mechanical device is usually used to move media units from storage into a drive where the data can be read or written. Generally it has safety properties similar to on-line storage.

1.2.3 Off-Line

Off-line storage requires some direct human action to provide access to the storage media. For example, inserting a tape into a tape drive or plugging in a cable. Because the data are not accessible via any computer except during limited periods in which they are written or read back, they are largely immune to a whole class of on-line backup failure modes. Access time will vary depending on whether the media are on-site or off-site.

1.2.4 Off-Site Data Protection

To protect against a disaster or other site-specific problem, many people choose to send backup media to an off-site vault. The vault can be as simple as a system administrator's home office or as sophisticated as a disaster-hardened, temperature-controlled, high-security bunker with facilities for backup media storage. Importantly a data replica can be off-site but also on-line (e.g., an off-site RAID mirror). Such a replica has fairly limited value as a backup, and should not be confused with an off-line backup.

1.2.5 Backup Site or Disaster Recovery Center (DR Center)

In the event of a disaster, the data on backup media will not be sufficient to recover. Computer systems onto which the data can be restored and properly configured networks are necessary too. Some organizations have their own data recovery centers that are equipped for this scenario. Other organizations contract this out to a third-party recovery center. Because a DR site is itself a huge investment, backing up is very rarely considered the preferred method of moving data to a DR site. A more typical way would be remote disk mirroring, which keeps the DR data as up to date as possible.

2. Selection and Extraction of Data

A successful backup job starts with selecting and extracting coherent units of data. Most data on modern computer systems is stored in discrete units, known as files. These files are organized into file systems. Files that are actively being updated can be thought of as "live" and present a challenge to back up. It is also useful to save metadata that describes the computer or the file system being backed up.

Deciding what to back up at any given time is a harder process than it seems to. By backing up too much redundant data, the data repository will fill up too quickly. Backing up an insufficient amount of data can eventually lead to the loss of critical information.

2.1 Files

2.1.1 Copying Files

With file-level approach, making copies of files is the simplest and most common way to perform a backup. A means to perform this basic function is included in all backup software and all operating systems.

2.1.2 Partial File Copying

Instead of copying whole files, one can limit the backup to only the blocks or bytes within a file that have changed in a given period of time. This technique can use substantially less storage space on the backup medium, but requires a high level of sophistication to reconstruct files in a restore situation. Some implementations require integration with the source file system.

2.1.3 Deleted Files

To prevent the unintentional restoration of files that have been intentionally deleted, a record of the deletion must be kept.

2.2 File Systems

2.2.1 File System Dump

Instead of copying files within a file system, a copy of the whole file system itself in block-level can be made. This is also known as a raw partition backup and is related to disk imaging. The process usually involves unmounting the file system and running a program like dd (Unix). Because the disk is read sequentially and with large buffers, this type of backup can be much faster than reading every file normally, especially when the file system contains many small files, is highly fragmented, or is nearly full. But because this method also reads the free disk blocks that contain no useful data, this method can also be slower than conventional reading, especially when the file system is nearly empty. Some file systems, such as XFS, provide a "dump" utility that reads the disk sequentially for high performance while skipping unused sections. The corresponding restore utility can selectively restore individual files or the entire volume at the operator's choice.

2.2.2 Identification of Changes

Some file systems have an archive bit for each file that says it has been recently changed. Some backup software looks at the date of the file and compares it with the last backup to determine whether the file has been changed.

2.2.3 Versioning File System

A versioning file system keeps track of all changes to a file and makes those changes accessible to the user. Generally this gives access to any previous version, all the way back to the file's creation time. An example of this is the Wayback versioning file system for Linux.

2.3 Live Data

If a computer system is in use while it is being backed up, the possibility of files being open for reading or writing is real. If a file is open, the contents on disk may not correctly represent what the owner of the file intends. This is especially true for database files of all kinds. The term fuzzy backup can be used to describe a backup of live data that looks like it runs correctly, but does not represent the state of the data at any single point in time. This is because the data being backed up changed in the period of time between when the backup started and when it finished.

Backup options for live (and other) data availability scenarios include:

2.3.1 Snapshot Backup

A snapshot is an instantaneous function of some storage systems that presents a copy of the file system as if it were frozen at a specific point in time, often by a copy-on-write mechanism. An effective way to back up live data is to temporarily quiesce them (e.g., close all files), take a snapshot, and then resume live operations. At this point the snapshot can be backed up through normal methods. While a snapshot is very handy for viewing a file system as it is at a different point in time, it is hardly an effective backup mechanism by itself.

2.3.2 Open File Backup

Many backup software packages feature the ability to handle open files in backup operations. Some simply check for openness and try again later. File locking is useful for regulating access to open files.

When attempting to understand the logistics of backing up open files, one must consider that the backup process could take several minutes to back up a large file such as a database. In order to back up a file that is in use, it is vital that the entire backup represent a single-moment snapshot of the file, rather than a simple copy of a read-through. This represents a challenge when backing up a file that is constantly changing. Either the database file must be locked to prevent changes, or a method must be implemented to ensure that the original snapshot is preserved long enough to be copied while changes are being preserved.

2.3.3 Cold Database (Offline) Backup

During a cold backup, the database is closed or locked and not available to users. The data files do not change during the backup process so the database is in a consistent state when it is returned to normal operation.

2.3.4 Hot Database (Online) Backup

Some database management systems offer a means to generate a backup image of the database while it is online and usable ("hot"). This usually includes an inconsistent image of the data files plus a log of changes made while the procedure is running. Upon a restore, the changes in the log files are reapplied to bring the copy of the database up-to-date (the point in time at which the initial hot backup ended).

2.4 Metadata

Not all information stored on the computer is stored in files. Accurately recovering a complete system from scratch requires keeping track of this non-file data too.

2.4.1 System Description

System specifications are needed to procure an exact replacement after a disaster.

2.4.2 Boot Sector

The boot sector can sometimes be recreated more easily than saving it. Still, it usually isn't a normal file and the system won't boot without it.

2.4.3 Partition Layout

The layout of the original disk, as well as partition tables and file system settings, is needed to properly recreate the original system.

2.4.4 File Metadata

Each files permissions, owner, group,and any other metadata need to be backed up for a restore to properly recreate the original environment.

2.4.5 System Metadata

Different operating systems have different ways of storing configuration information. Microsoft Windows keep a registry of system information that is more difficult to restore than a typical file.

New Words

file	[faɪl]	*n.* 文件
restore	[rɪˈstɔː]	*vt.&vi.* 恢复
recover	[rɪˈkʌvə]	*vt.* 恢复；重新获得；找回
deletion	[dɪˈliːʃn]	*n.* 删除
corruption	[kəˈrʌpʃn]	*n.* 毁坏
reconstitute	[ˌriːˈkɒnstɪtjuːt]	*vt.* 再组成，再构成
configuration	[kənˌfɪgəˈreɪʃn]	*n.* 布局，构造；配置
significant	[sɪgˈnɪfɪkənt]	*adj.* 重要的；显著的；有重大意义的
structure	[ˈstrʌktʃə]	*n.* 结构；构造；体系 *vt.* 构成，排列

portability	[ˌpɔtəˈbɪlɪtɪ]	*n.* 可携带，轻便
optimize	[ˈɒptɪmaɪz]	*vt.* 使最优化
optimization	[ˌɒptɪmaɪˈzeɪʃən]	*n.* 最佳化，最优化
validate	[ˈvælɪdeɪt]	*vt.* 批准，确认；证实
reliability	[rɪˌlaɪəˈbɪlɪtɪ]	*n.* 可靠，可靠性，可信赖
setup	[ˈsetʌp]	*n.* 计划
recoverability	[rɪˌkʌvərəˈbɪlətɪ]	*n.* 可复（原）性
incremental	[ˌɪŋkrəˈmentl]	*adj.* 增加的
feasible	[ˈfiːzəbl]	*adj.* 可行的；可用的；可实行的；可能的
eliminate	[ɪˈlɪmɪneɪt]	*vt.* 排除，消除
unchanged	[ʌnˈtʃeɪndʒd]	*adj.* 未改变的，无变化的
infrequent	[ɪnˈfrɪkwənt]	*adj.* 不频繁的
interval	[ˈɪntəvl]	*n.* 间隔
successive	[səkˈsesɪv]	*adj.* 连续的，相继的
synthesize	[ˈsɪnθɪsaɪz]	*v.* 合成，综合
characteristic	[ˌkærəktəˈrɪstɪk]	*adj.* 特有的；独特的；表示特性的 *n.* 特性，特征，特色
protection	[prəˈtekʃn]	*n.* 保护
host	[həʊst]	*n.* 主机
byte	[baɪt]	*n.* 字节
restoration	[ˌrestəˈreɪʃn]	*n.* 恢复，复原
balance	[ˈbæləns]	*n.* 平衡
accessibility	[əkˌsesəˈbɪlɪtɪ]	*n.* 可访问性，可存取性
exclusive	[ɪkˈskluːsɪv]	*adj.* 专用的；单独的 *n.* 专有物
millisecond	[ˈmɪlɪsekənd]	*n.* 毫秒
overwrite	[ˌəʊvəˈraɪt]	*v.* 改写
payload	[ˈpeɪləʊd]	*n.*（炸弹、导弹的）爆炸力
disaster	[dɪˈzɑːstə]	*n.* 灾难；彻底的失败
bunker	[ˈbʌŋkə]	*n.* 地堡
replica	[ˈreplɪkə]	*n.* 复制品
scenario	[səˈnɑːrɪəʊ]	*n.* 可能发生的情况；设想
coherent	[kəʊˈhɪərənt]	*adj.* 一致的；连贯的；条理分明的；清楚明白的
insufficient	[ˌɪnsəˈfɪʃnt]	*adj.* 不足的，不够的

partial	[ˈpɑːʃl]	*adj.* 部分的
block	[blɒk]	*n.* 块
		vt. 阻止；阻塞；限制
substantially	[səbˈstænʃəlɪ]	*adv.* 本质上，实质上；充分地
sophistication	[səˌfɪstɪˈkeɪʃn]	*n.* 老练，精明
reconstruct	[ˌriːkənˈstrʌkt]	*vt.* 重建；重现；改造
unintentional	[ˌʌnɪnˈtenʃənl]	*adj.* 不是故意的；无意的，无心的
intentionally	[ɪnˈtenʃənəlɪ]	*adv.* 有意地，故意地
dump	[dʌmp]	*vt.* 转储，转贮
corresponding	[ˌkɒrəˈspɒndɪŋ]	*adj.* 相当的，对应的；符合的
identification	[aɪˌdentɪfɪˈkeɪʃn]	*n.* 鉴定，识别；验明；身份证明
instantaneous	[ˌɪnstənˈteɪnɪəs]	*adj.* 瞬间的；即刻的
quiesce	[kwaɪˈes]	*v.* 静默，沉寂
regulate	[ˈregjʊleɪt]	*v.* 控制；管理；调节，调整
constantly	[ˈkɒnstəntlɪ]	*adv.* 不断地，时常地；时刻
inconsistent	[ˌɪnkənˈsɪstənt]	*adj.* 不一致的，不调和的
recreate	[ˌriːkrɪˈeɪt]	*v.* 重建；再创造
boot	[buːt]	*vt.* 引导
permission	[pəˈmɪʃn]	*n.* 允许；批准，正式认可

Phrases

data deletion	数据删除
data repository model	数据仓库模型
storage location	存储位置
deal with	处理
dry run	演习，排练
backup scheme	备份方案
backup rotation scheme	备份计划
diverse system	多样化系统
incremental style repository	增量样式存储库
full backup	全备份
differential backup	差异备份
reverse delta type repository	反向增量类型存储库
hard link	硬链接

disk mirroring 磁盘镜像
disk array 磁盘阵列
tape library 磁带库
operating system 操作系统
raw partition backup 原始分区备份
disk imaging 磁盘映像
fuzzy backup 模糊备份
snapshot backup 快照备份
copy-on-write mechanism 写时复制机制
software package 软件包
file locking 文件锁定
cold backup 冷备份
database management system 数据库管理系统
log file 日志文件
boot sector 引导扇区

Abbreviations

PC（Personal Computer） 个人计算机
SAN（Storage Area Network） 存储区域网络
DR（Disaster Recovery） 灾难恢复

Exercises

【Ex.5】Answer the following questions according to the text.

1. What does a data backup, or the process of backing up, refer to in information technology?
2. What are the purposes of backups?
3. What does an incremental style repository aim to do?
4. What does a reverse delta type repository do?
5. What do data repository implementations include?
6. What will happen by backing up too much redundant data? What about backing up an insufficient amount of data?
7. What is a raw partition backup? What is it related to?
8. What is a snapshot?
9. What is an effective way to back up live data?
10. What happens to the database during a cold backup?

参考译文

提取、转换、加载

ETL 是一种数据集成，指用于整合来自多源头数据的三个步骤（提取、转换、加载）。它通常用于构建数据仓库。在此过程中，从源系统获取（提取）数据，将其改变（转换）为可以分析的格式，并存储（加载）到数据仓库或其他系统中。提取、加载、转换（ELT）是一种替代而不是相关的方法，旨在将处理推送到数据库以提高性能。

1. 为什么 ETL 很重要

多年来，企业依靠 ETL 流程来获得数据的整合视图，以便做出更好的商业决策。如今，这种集成来自多个系统和源头的数据的方法仍然是组织数据集成工具箱中的核心组件。

- 与企业数据仓库（静态数据）一起使用时，ETL 为业务提供深入的历史背景。
- 通过提供整合视图，ETL 使业务用户可以更轻松地分析与其计划相关的数据并形成报告。
- ETL 可以提高数据专业人员的工作效率，因为它编写并重用了移动数据的流程，而无需编写代码或脚本的技术技能。
- 随着时间的推移，ETL 不断发展，它可以支持像流数据这样的新集成需求。
- 组织需要 ETL 和 ELT 将数据整合在一起，保持准确性并提供所需的审计，以便把数据入库，对数据进行分析和形成报告。

2. 如何使用 ETL

核心 ETL 和 ELT 工具与其他数据集成工具以及数据管理的其他方面（例如数据质量、数据治理、虚拟化和元数据）协同工作，目前广泛应用于以下几个方面。

2.1 ETL 和传统用途

ETL 是一种经过验证的方法，许多组织每天都依赖这种方法——如需要定期查看销售数据的零售商，或寻求准确描述索赔的医疗服务提供者。ETL 可以组合并显示来自仓库或其他数据存储的交易数据，以便业务人员以他们能够理解的格式查看。ETL 还用于将数据从老旧系统迁移到具有不同数据格式的现代系统，这通常用于整合来自业务合并的数据，以及那些外部供应商或合作伙伴收集和加入的数据。

2.2 大数据的 ETL——转换和适配器

获得最多数据的人获胜。虽然这不一定正确，但轻松访问广泛的数据可以为企业提供竞争优势。如今，企业需要访问各种大数据——来自视频、社交媒体、物联网（IoT）、服务器日志的数据，以及空间数据、开放或众包数据等。ETL 供应商经常为其工具添加新变化，以支持这些新兴需求和新数据源。适配器允许访问各种各样的数据源，数据集成工具与这些适配器交互以便有效地提取和加载数据。

2.3　用于 Hadoop 的 ETL——以及更多

ETL 已经发展到支持集成，而不仅仅是传统的数据仓库。高级 ETL 工具可以将结构化和非结构化数据加载并转换到 Hadoop 中。这些工具从 Hadoop 并行读取和写入多个文件，简化了数据合并到公共转换的过程。一些解决方案包含针对在 Hadoop 上运行的事务和交互数据的预构建 ETL 转换库。ETL 还可以与跨事务系统、运营数据存储、BI 平台、主数据管理（MDM）中心和云相集成。

2.4　ETL 和自助数据访问

自助数据准备是一种快速发展的趋势，它将访问、混合和转换数据的能力交付给业务用户和其他非技术数据专业人员。这种方法可提高组织敏捷性，使 IT 免于为业务用户以不同格式配置数据的负担，并使花费在数据准备上的时间更少，花费在洞察本质上的时间更多。因此，业务和 IT 数据专业人员都可以提高工作效率，组织也可以扩展数据使用以做出更好的决策。

2.5　ETL 和数据质量

ETL 和其他数据集成软件工具——用于数据清理、分析和审计——确保数据值得信赖。ETL 工具能与数据质量工具集成，ETL 供应商在其解决方案中包含了相关工具（例如用于数据映射和数据沿袭的工具）。

2.6　ETL 和元数据

元数据有助于我们了解数据的沿袭（来自何处）及其对组织中其他数据资产的影响。随着数据架构变得越来越复杂，跟踪组织中不同数据元素的使用和相关性非常重要。例如，如果将 Twitter 账户名添加到客户数据库，则需要明确会对哪些产生影响，如对 ETL 作业、应用程序或报告的影响。

3. ETL 如何工作

ETL 与许多其他数据集成功能、流程和技术密切相关。

3.1　SQL

结构化查询语言是访问和转换数据库中数据的最常用方法。

3.2　转换、业务规则和适配器

提取数据后，ETL 根据业务规则将数据转换为新格式。然后将转换后的数据加载到目标中。

3.3　数据映射

数据映射是转换过程的一部分。映射为应用程序提供了如何获取其需要处理的数据的详细说明。它还描述了哪个源字段映射到哪个目标字段。例如，网站活动的数据馈送中的第三个属性可能是用户名，第四个可能是该活动发生时的时间戳，第五个可能是用户点击的产品。使用该数据的应用程序或 ETL 过程必须将来自源系统的这些相同字段或属性（即，网站活动数据馈送）映射为目标系统所需的格式。如果目标系统是客户关系管理系统，它可能首先存储用户名，将存储时间戳排到第五，它可能根本不存储所选产品。在这种情况下，在从源读取数据并将数据写入目标之间，可能会发生以预期格式（以正确顺序）进行日期格式转换。

3.4 脚本

ETL 是一种自动化脚本（指令集）的方法，这些脚本在幕后运行以移动和转换数据。在 ETL 之前，脚本分别用 C 或 COBOL 编写，以在特定系统之间传输数据，这导致了多个数据库运行大量脚本。早期的 ETL 工具以批处理方式运行在大型机上，ETL 后来迁移到 UNIX 和 PC 平台。今天的组织仍然使用脚本和程序化数据移动方法。

3.5 ETL 与 ELT

早先只有 ETL。后来，组织增加了 ELT，将它作为一种补充方法。ELT 从源系统中提取数据，将其加载到目标系统，然后使用源系统的处理能力进行转换。这加速了数据处理，因为它发生在数据所在之处。

3.6 数据质量

在集成数据之前，通常会创建一个临时区域，以便清理数据，数据值可以标准化（NC 和 North Carolina，Mister 和 Mr.，或 Matt 和 Matthew），可以验证地址并删除重复项。许多解决方案仍然是独立的，但数据质量程序现在可以作为数据集成过程中转换的一部分来运行。

3.7 调度和处理

ETL 工具和技术可以提供批量调度或实时功能。还可以在服务器中处理高容量数据，也可将处理推迟到数据库级别。与专用引擎相比，这种在数据库中处理的方法避免了数据重复，并且防止了在数据库平台上需要添加的额外容量。

3.8 批处理

ETL 通常指在所谓的“批处理窗口”期间在两个系统之间移动大量数据的批处理过程。在此设定的时间段内——如中午和下午 1 点之间——数据同步时，源系统或目标系统不会发生任何操作。大多数银行都会进行夜间批处理以解决全天发生的交易。

3.9 网络服务

网络服务是一种基于互联网的方法，可以近乎实时地向各种应用程序提供数据或功能。此方法简化了数据集成过程，可以更快地从数据中提供更多价值。例如，假设客户联系你的呼叫中心。你可以创建一个网络服务，只需将电话号码传递给从多个源或 MDM 中心提取数据的网络服务，即可在亚秒内返回完整的客户配置文件。通过更丰富的客户知识，客户服务代表可以更好地决定如何与客户进行交互。

3.10 主数据管理（MDM）

MDM 是将数据放到一起以创建跨多个源的单个数据视图的过程。它包括 ETL 和数据集成功能，将数据混合在一起并创建“黄金记录”或“最佳记录”。

3.11 数据虚拟化

虚拟化是一种灵活的方法，可以将数据混合在一起，以创建数据的虚拟视图，而无需移动数据。数据虚拟化与 ETL 不同，因为即使仍然发生映射和连接数据，也不需要物理临时表来存储结果。这是因为视图通常存储在内存中并缓存以提高性能。某些数据虚拟化解决方案（如 SAS Federation Server）提供动态数据屏蔽、随机化和散列函数，以保护敏感数据免受特定角色或组的影响。SAS 还在生成视图时提供按需数据质量管理。

3.12 事件流处理和 ETL

当数据速度增加到每秒数百万个事件时，事件流处理可用于监视数据流、处理数据流并帮助做出更及时的决策。

Unit 5

Text A
Python Programming Language

Python features a dynamic type system and automatic memory management. It supports multiple programming paradigms, including object-oriented, imperative, functional and procedural, and has a large and comprehensive standard library.

1. Features and Philosophy

Python is a multi-paradigm programming language. Object-oriented programming and structured programming are fully supported, and many of its features support functional programming and aspect-oriented programming. Many other paradigms are supported via extensions, including design by contract and logic programming.

Python uses dynamic typing, and a combination of reference counting and a cycle-detecting garbage collector for memory management. It also features dynamic name resolution (late binding), which binds method and variable names during program execution.

2. Syntax and Semantics

Python is meant to be an easily readable language. Its formatting is visually uncluttered, and it often uses English keywords where other languages use punctuation. Unlike many other languages, it does not use curly brackets to delimit blocks, and semicolons after statements are optional. It has fewer syntactic exceptions and special cases than C or Pascal.

2.1 Indentation

Python uses whitespace indentation, rather than curly brackets or keywords, to delimit blocks. An increase in indentation comes after certain statements; a decrease in indentation signifies the end of the current block. Thus, the program's visual structure accurately represents the program's semantic structure. This feature is also sometimes termed the off-side rule.

2.2 Statements and Control Flow

Python's statements include:

- The assignment statement (token "=", the equals sign).
- The if statement, which conditionally executes a block of code, along with else and elif (a contraction of else-if).
- The for statement, which iterates over an iterable object, capturing each element to a local variable for use by the attached block.
- The while statement, which executes a block of code as long as its condition is true.
- The raise statement, used to raise a specified exception or re-raise a caught exception.
- The class statement, which executes a block of code and attaches its local namespace to a class, for use in object-oriented programming.
- The def statement, which defines a function or method.
- The pass statement, which serves as a NOP. It is syntactically needed to create an empty code block.
- The assert statement, used during debugging to check for conditions that ought to apply.
- The yield statement, which returns a value from a generator function. From Python 2.5, yield is also an operator.
- The import statement, which is used to import modules whose functions or variables can be used in the current program. There are three ways of using import: import <module name> [as <alias>] or from <module name> import * or from <module name> import <definition 1> [as <alias 1>], <definition 2> [as <alias 2>], ...
- The print statement was changed to the print() function in Python 3.

2.3 Expressions

Some Python expressions are similar to languages such as C and Java, while some are not.

- Addition, subtraction, and multiplication are the same, but the behavior of division differs. There are two types of divisions in Python. Python also adds the ** operator for exponentiation.
- From Python 3.5, the new @ infix operator is introduced. It is intended to be used by libraries such as NumPy for matrix multiplication.
- In Python, == compares by value, versus Java, which compares numerics by value and objects by reference. Python's "is" operator may be used to compare object identities. In Python, comparisons may be chained, for example a <= b <= c.
- Python uses the words and, or, not for its boolean operators rather than the symbolic &&, ||, ! used in Java and C.
- Python has a type of expression termed a list comprehension. Python 2.4 extended list comprehensions into a more general expression termed a generator expression.
- Conditional expressions in Python are written as x if c else y (different in order of operands

from the c ? x : y operator common to many other languages).

- Python makes a distinction between lists and tuples. Lists are written as [1, 2, 3], are mutable, and cannot be used as the keys of dictionaries (dictionary keys must be immutable in Python). Tuples are written as (1, 2, 3), which are immutable and thus can be used as the keys of dictionaries, provided all elements of the tuple are immutable.
- Python has a "string format" operator %. This functions analogously to printf format strings in C, e.g. "spam=%s eggs=%d" % ("blah", 2).
- Python has various kinds of string literals:

① Strings delimited by single or double quote marks. Both kinds of strings use the backslash (\) as an escape character.

② Triple-quoted strings, which begin and end with a series of three single or double quote marks.

③ Raw string, denoted by prefixing the string literal with an r. Escape sequences are not interpreted.

3. Libraries

Python's large standard library, commonly cited as one of its greatest strengths, provides tools suited to many tasks. For Internet-facing applications, many standard formats and protocols such as MIME and HTTP are supported. It includes modules for creating graphical user interfaces, connecting to relational databases, generating pseudorandom numbers, arithmetic with arbitrary precision decimals, manipulating regular expressions, and unit testing.

Some parts of the standard library are covered by specifications, but most modules are not. They are specified by their code, internal documentation, and test suites (if supplied). However, because most of the standard library is cross-platform Python code, only a few modules need altering or rewriting.

As of March 2018, the Python Package Index (PyPI), the official repository for third-party Python software, contains over 130,000 packages with a wide range of functionality, including:

- Graphical user interfaces
- Web frameworks
- Multimedia
- Databases
- Networking
- Test frameworks
- Automation
- Web scraping
- Documentation
- System administration
- Scientific computing

- Text processing
- Image processing

4. Development Environments

Most Python implementations (including CPython) include a read-eval-print loop (REPL), permitting them to function as a command line interpreter for which the user enters statements sequentially and receives results immediately.

Other shells add further abilities such as auto-completion, session state retention and syntax highlighting.

As well as standard desktop integrated development environments, there are Web browser-based IDEs: SageMath, intended for developing science and math-related Python programs; PythonAnywhere, a browser-based IDE and hosting environment; and Canopy IDE, a commercial Python IDE emphasizing scientific computing.

New Words

feature	[ˈfiːtʃə]	*n.* 特征，特点
		vt. 使有特色；描写……的特征
dynamic	[daɪˈnæmɪk]	*adj.* 动态的
paradigm	[ˈpærədaɪm]	*n.* 范例，样式
object-oriented	[ˈɒbdʒɪkt-ˈɔːrɪəntɪd]	*adj.* 面向对象的
imperative	[ɪmˈperətɪv]	*adj.* 命令的
		n. 命令，规则
functional	[ˈfʌŋkʃənl]	*adj.* 功能的；函数的
procedural	[prəˈsiːdʒərəl]	*adj.* 程序的，过程的
philosophy	[fəˈlɒsəfɪ]	*n.* 哲学；哲学体系
metaprogramming	[ˌmetəpˈrəʊgræmɪŋ]	*adj.* 元程序设计的，元编程的
metaobject	[ˌmetəˈɒbdʒɪkt]	*adj.* 元对象的
logic	[ˈlɒdʒɪk]	*n.* 逻辑，逻辑学
		adj. 逻辑的
resolution	[ˌrezəˈluːʃn]	*n.* 解析；分辨率
binding	[ˈbaɪndɪŋ]	*n.* 绑定
		adj. 捆绑的
		v. 绑定
variable	[ˈveərɪəbl]	*n.* 变量
		adj. 变量的
syntax	[ˈsɪntæks]	*n.* 语法，句法

semantic	[sɪ'mæntɪk]	*adj.* 语义的
readable	['riːdəbl]	*adj.* 易读的；易懂的
uncluttered	[ˌʌn'klʌtəd]	*adj.* 整齐的，整洁的
punctuation	[ˌpʌŋktʃu'eɪʃn]	*n.* 标点法；标点符号
semicolon	[ˌsemɪ'kəʊlən]	*n.* 分号
statement	['steɪtmənt]	*n.* 声明；语句
optional	['ɒpʃənl]	*adj.* 可选择的；任意的；非强制的
indentation	[ˌɪnden'teɪʃn]	*n.* 缩进，缩格
delimit	[di'lɪmɪt]	*vt.* 界定；限制
decrease	[dɪ'kriːs]	*n.* 减少，减小；减少量 *v.* 减少，减小
signify	['sɪgnɪfaɪ]	*vt.* 表示……的意思；意味；预示 *vi.* 具有重要性，要紧
accurately	['ækjʊrətlɪ]	*adv.* 正确无误地，准确地；精确地
represent	[ˌreprɪ'zent]	*v.* 代表
conditionally	[kən'dɪʃənəlɪ]	*adv.* 有条件地
execute	['eksɪkjuːt]	*vt.* 执行
iterate	['ɪtəreɪt]	*vt.* 迭代，重复
capture	['kæptʃə]	*vt.* & *n.* 捕获
element	['elɪmənt]	*n.* 元素，要素
attached	[ə'tætʃt]	*adj.* 附加的，附属的
condition	[kən'dɪʃn]	*n.* 条件；状态
exception	[ɪk'sepʃn]	*n.* 异常；例外，除外
namespace	['neɪmspeɪs]	*n.* 命名空间，名字空间
syntactically	[sɪn'tæktɪklɪ]	*adv.* 依照句法地，在语句构成上
create	[kri'eɪt]	*vt.* 建立，创建
empty	['emptɪ]	*adj.* 空的
debug	[ˌdiː'bʌg]	*vt.* 调试程序，排除故障
return	[rɪ'tɜːn]	*v.* 返回
generator	['dʒenəreɪtə]	*n.* 生成器
infix	['ɪnfɪks]	*vt.* 让……插进 *n.* 插入词，中缀
boolean	['buːlɪən]	*adj.* 布尔的
symbolic	[sɪm'bɒlɪk]	*adj.* 符号的；象征性的
comprehension	[ˌkɒmprɪ'henʃn]	*n.* 理解；包含
operand	['ɒpərænd]	*n.* 操作数；运算数
tuple	[tʌpl]	*n.* 元组

mutable	[ˈmjuːtəbl]	*adj.* 可变的，易变的
analogous	[əˈnæləgəs]	*adj.* 相似的，可比拟的
backslash	[ˈbækslæʃ]	*n.* 反斜线符号
interpret	[ɪnˈtɜːprɪt]	*vt.* 解释 *vi.* 作解释
library	[ˈlaɪbrərɪ]	*n.* 库
standard	[ˈstændəd]	*n.* 标准，规格 *adj.* 标准的
strength	[streŋθ]	*n.* 力量；优点，长处
task	[tɑːsk]	*n.* 工作，任务；作业
protocol	[ˈprəʊtəkɒl]	*n.* 协议
connect	[kəˈnekt]	*vt.* 连接，联结 *vi.* 连接；建立关系
pseudorandom	[psjuːdəʊˈrændəm]	*adj.* 伪随机的
arithmetic	[əˈrɪθmətɪk]	*n.* 算法
arbitrary	[ˈɑːbɪtrərɪ]	*adj.* 任意的，随意的
manipulate	[məˈnɪpjʊleɪt]	*vt.* 操作，处理
cross-platform	[krɔsˈplætfɔːm]	*n.* 跨平台
rewrite	[ˌriːˈraɪt]	*vt.* 重写，改写
third-party	[ˈθɜːdpaːtɪ]	*adj.* 第三方的
package	[ˈpækɪdʒ]	*n.* 软件包
framework	[ˈfreɪmwɜːk]	*n.* 构架；框架
multimedia	[ˌmʌltɪˈmiːdɪə]	*n.* 多媒体 *adj.* 多媒体的
networking	[ˈnetwɜːkɪŋ]	*n.* 网络化
automation	[ˌɔːtəˈmeɪʃn]	*n.* 自动化（技术），自动操作
scrape	[skreɪp]	*v.* 抓取
documentation	[ˌdɒkjʊmenˈteɪʃn]	*n.* 文档
session	[ˈseʃn]	*n.* 会话
retention	[rɪˈtenʃn]	*n.* 保留
emphasize	[ˈemfəsaɪz]	*vt.* 强调，着重；使突出

Phrases

automatic memory management	自动内存管理
multiple programming paradigm	多程序设计范式

standard library	标准库
multi-paradigm programming language	多范式编程语言
structured programming	结构化程序设计
aspect-oriented programming	面向方面程序设计
design by contract	契约式设计
a combination of ...	……的组合
cycle-detecting garbage collector	循环检测垃圾收集器
memory management	内存管理
dynamic name resolution	动态名称解析
variable name	变量名
curly bracket	波形括号；大括号
off-side rule	缩排规则
control flow	控制流
assignment statement	赋值语句
local variable	局部变量
as long as	只要；如果
object-oriented programming	面向对象程序设计
code block	代码块
generator function	生成器函数
infix operator	插入算符，中缀运算符
list comprehension	列表推导式
generator expression	生成器表达式
conditional expression	条件表达式
string format	字符串格式
quote mark	引号
escape character	转义字符
a series of	一系列；一连串
raw string	原始字符串
Internet-facing application	面向互联网的应用
graphical user interface	图形用户界面
pseudorandom number	伪随机数
regular expression	正则表达式
unit testing	单元测试
test suite	测试套件
image processing	图像加工，图像处理
command line interpreter	命令行解释程序
session state retention	会话状态保留

Abbreviations

NOP（NoOperation） 无操作
MIME（Multipurpose Internet Mail Extensions） 多用途互联网邮件扩展
HTTP（HyperText Transfer Protocol） 超文本传输协议
REPL（read-eval-print loop） 读取—求值—输出循环
IDE（Integrated Development Environment） 集成开发环境

Exercises

【Ex.1】Fill in the following blanks according to the text.

1. Python features ________ and ________. It supports multiple programming paradigms, including ________, ________, ________ and procedural, and has a large and comprehensive ________.

2. Python uses ________, and a combination of ________ and ________ for memory management.

3. Python is meant to be ________. Its formatting is ________, and it often uses ________where other languages use punctuation.

4. Python uses ________, rather than curly brackets or ________, to delimit blocks. An increase in indentation comes after ________; a decrease in indentation signifies ________.

5. The for statement iterates over ________, capturing each element to a local variable for use by ________.

6. The import statement is used to ________ whose functions or ________ can be used in the current program. There are ________ ways of using import.

7. Some Python expressions are ________ languages such as C and Java, while some are not. Python uses the words and, or, not for ________ rather than the symbolic &&, ||, ! used in ________. Python has a type of expression termed ________.

8. Python makes a distinction between ________ and ________. Lists are written as [1, 2, 3], are ________, and cannot be used as ________. Tuples are written as ________, are immutable and thus can be used as the keys of dictionaries, provided ________.

9. Some parts of the standard library are covered by ________, but most modules are not. They are specified by ________, ________, and ________.

10. Most Python implementations (including CPython) include ________, permitting them to function as ________for which the user enters statements sequentially and ________.

【Ex.2】Translate the following terms or phrases from English into Chinese and vice versa.

1. code block　　1. ______
2. variable name　　2. ______
3. control flow　　3. ______
4. graphical user interface　　4. ______
5. memory management　　5. ______
6. *vt.*调试程序，排除故障　　6. ______
7. 正则表达式　　7. ______
8. 单元测试　　8. ______
9. *n.*算法　　9. ______
10. *n.*赋值；分配　　10. ______

【Ex.3】Translate the following passages into Chinese.

OOP

Object-oriented programming (OOP) is a programming language model in which programs are organized around data, or objects, rather than functions and logic. An object can be defined as a data field that has unique attributes and behavior. Examples of an object can range from physical entities, such as a human being that is described by properties like name and address.

The first step in OOP is to identify all of the objects a programmer wants to manipulate and how they relate to each other, an exercise often known as data modeling. Once an object is known, it is generalized as a class of objects that defines the kind of data it contains and any logic sequences that can manipulate it. Each distinct logic sequence is known as a method and objects can communicate with well-defined interfaces called messages.

Simply put, OOP focuses on the objects that developers want to manipulate rather than the logic required to manipulate them. This approach to programming is well-suited for programs that are large, complex and actively updated or maintained. Due to the organization of an object-oriented program, this method is also conducive to collaborative development where projects can be divided into groups. Additional benefits of OOP include code reusability, scalability and efficiency.

【Ex.4】Fill in the blanks with the words given below.

languages	instructions	interchangeably	behavior	abstract
computation	describe	subset	emphasize	specification

Programming Language

A computer programming language is a language used to write computer programs, which involves a computer performing some kind of ___1___ or algorithm and possibly controling external devices such asprinters, disk drives, robots, and so on. For example, PostScript programs are frequently created by another program to control a computer printer or display. More generally, a programming language may ____2____ computation on some, possibly abstract, machine. It is generally accepted that a complete specification for a programming language includes a description of a machine or processor for that language. Programming ____3____ differ from natural languages in that natural languages are only used for interactions between people, while programming languages also allow humans to communicate ____4____ to machines.

The term computer language is sometimes used ____5____ with programming language. However, the usage of both terms varies among authors, including the exact scope of each. One usage describes programming languages as a ____6____ of computer languages. In this vein, languages used in computing that have a different goal than expressing computer programs are generically designated computer languages. For instance, markup languages are sometimes referred to as computer languages to ____7____ that they are not meant to be used for programming.

Another usage regards programming languages as theoretical constructs for programming ____8____ machines, and computer languages as the subset that runs on physical computers, which have finite hardware resources. John C. Reynolds emphasizes that formal ____9____ languages are just as much programming languages as are the languages intended for execution. He also argues that textual and even graphical input formats that affect the ____10____ of a computer are programming languages, despite the fact they are commonly not Turing-complete, and remarks that ignorance of programming language concepts is the reason for many flaws in input formats.

Text B
R Programming Language

扫码听课文

1. Introduction

R is more than a programming language. It is an interactive environment for doing statistics. I find it more helpful to think of R as having a programming language than being a programming language. The R language is the scripting language for the R environment, just as VBA is the scripting language for Microsoft Excel. Some of the more unusual features of the R language begin to make sense when viewed from this perspective.

2. Assignment and Underscore

The assignment operator in R is <- as in

```
e <- m*c^2.
```

It is also possible, though uncommon, to reverse the arrow and put the receiving variable on the right, as in

```
m*c^2->e.
```

It is sometimes possible to use = for assignment, though I don't understand when this is and is not allowed. Most people avoid the issue by always using the arrow.

However, when supplying default function arguments or calling functions with named arguments, you must use the = operator and cannot use the arrow.

At some time in the past R, or its ancestor S, used underscore as assignment. This meant that the C convention of using underscores as separators in multi-word variable names was not only disallowed but produced strange side effects. For example, first_name would not be a single variable but rather the instruction to assign the value of name to the variable first! S-PLUS still follows this use of the underscore. However, R allows underscore as a variable character and not as an assignment operator.

3. Vectors

The primary data type in R is the vector. Before describing how vectors work in R, it is helpful to distinguish two ideas of vectors in order to set the correct expectations.

The first idea of a vector is what I will call a container vector. This is an ordered collection of numbers with no other structure, such as the vector<> container in C++. The length of a vector is the number of elements in the container. Operations are applied componentwise. For example, given two vectors x and y of equal length, $x*y$ would be the vector whose nth component is the product of the nth components of x and y. Also, log(x) would be the vector whose nth component is the logarithm of the nth component of x.

The other idea of a vector is a mathematical vector, an element of a vector space. In this context "length" means geometrical length determined by an inner product; the number of components is called "dimension." In general, operations are not applied componentwise. The expression $x*y$ is a single number, the inner product of the vectors. The expression log(x) is meaningless.

A vector in R is a container vector, a statistician's collection of data, not a mathematical vector. The R language is designed around the assumption that a vector is an ordered set of measurements rather than a geometrical position or a physical state (R supports mathematical vector operations, but they are secondary in the design of the language).

Adding a vector of length 22 and a vector of length 45 in most languages would raise an exception; the language designers would assume the programmer has made an error and the program is now in an undefined state. However, R allows adding two vectors regardless of their relative lengths. The elements of the shorter summand are recycled as often as necessary to create a vector the length of the longer summand. This is not attempting to add physical vectors that are incompatible for addition, but rather a syntactic convenience for manipulating sets of data (R does issue a warning when adding vectors of different lengths and the length of the longer vector is not an integer multiple of the length of the shorter vector. So, for example, adding vectors of lengths 3 and 7 would cause a warning, but adding vectors of length 3 and 6 would not).

The R language has no provision for scalars, nothing like a double in C-family languages. The only way to represent a single number in a variable is to use a vector of length one. And while it is possible to iterate through vectors as one might do in a for loop in C, it is usually clearer and more efficient in R to operate on vectors as a whole.

Vectors are created using the c function. For example, p <- c(2,3,5,7) sets p to the vector containing the first four prime numbers.

Vectors in R are indexed starting with 1 and matrices in are stored in column-major order. In both of these ways R resembles FORTRAN.

Elements of a vector can be accessed using[]. So in the above example,p[3] is 5.

Vectors automatically expand when assigning to an index past the end of the vector, as in Perl.

Boolean values can also be used as indices, and they behave differently than integers.

4. Sequences

The expression seq(a, b, n) creates a closed interval from a to b in steps of size n. For example,seq(1, 10, 3) returns the vector containing 1, 4, 7, and 10. This is similar to range(a, b, n) in Python, except that Python uses half-open intervals and so the 10 would not be included in this example. The step size argument defaults to 1 in both R and Python.

The notation a:b is an abbreviation for seq(a, b, 1).

The notation seq(a, b, length=n) is a variation that will set the step size to (b-a)/(n-1) so that the sequence has n points.

5. Types

The type of a vector is the type of the elements it contains and must be one of the following: Logical, integer, double, complex, character, or raw. All elements of a vector must have the same underlying type. This restriction does not apply to lists.

Type conversion functions have the naming convention as.xxx for the function converts its argument to type xxx. For example,as.integer(3.2) returns the integer 3, and as.character(3.2) returns the string “3.2”.

6. Boolean Operators

You can input T or TRUE for true values and F or FALSE for false values.

The operators & and | apply element-wise on vectors. The operators && and || are often used in conditional statements and use lazy evaluation as in C: The operators will not evaluate their second argument if the return value is determined by the first argument.

7. Lists

Lists are like vectors, except that elements need not all have the same type. For example, the first element of a list could be an integer and the second element be a string or a vector of Boolean values.

Lists are created using the list function. Elements can be accessed by position using [[]]. Named elements may be accessed either by position or by name.

Named elements of lists act like C structs, except that a dollar sign rather than a dot is used to access elements. For example, consider,

```
a <- list(name="Joe", 4, foo=c(3,8,9))
```

Now a[[1]] and a$name both equal the string "Joe".

8. Matrices

In a sense, R does not support matrices, only vectors. But you can change the dimension of a vector, essentially making it a matrix.

For example, m <- array(c(1,2,3,4,5,6), dim=c(2,3)) creates a matrix m. However, it may come as a surprise that the first row of m has elements 1, 3, and 5. This is because by default, R fills matrices by column, like FORTRAN. To fill m by row, add the argument by.row = TRUE to the call to the array function.

9. Comments

Comments begin with # and continue to the end of the line, as in Python or Perl.

10. Functions

The function definition syntax of R is similar to that of JavaScript. For example:

```
f <- function(a, b)
{
    return (a+b)
}
```

The function returns a function, which is usually assigned to a variable, f in this case, but need not be. You may use the function statement to create an anonymous function.

Note that return is a function; its argument must be contained in parentheses, unlike C where parentheses are optional. The use of return is optional; otherwise the value of the last line executed in a function is its return value.

Default values are defined similarly to C++. In the following example,b is set to 10 by default.

```
f <- function(a, b=10)
{
    return (a+b)
}
```

So f(5, 1) would return 6, and f(5) would return 15. R allows more sophisticated default values than does C++. A default value in R need not be a static type but could, for example, be a function of other arguments.

Function parameters are passed by value. The most common mechanism for passing variables by reference is to use non-local variables. (Not necessarily global variables, but variables in the calling routine's scope.) A safer alternative is to explicitly pass in all needed values and return a list as output.

New Words

programming	[ˈprəʊgræmɪŋ]	*n.* 编程
interactive	[ˌɪntərˈæktɪv]	*adj.* 交互式的，互动的
statistic	[stəˈtɪstɪk]	*n.* 统计，统计论 *adj.* 统计（上）的，统计学（上）的
perspective	[pəˈspektɪv]	*n.* 观点，看法
assignment	[əˈsaɪnmənt]	*n.* 赋值
underscore	[ˌʌndəˈskɔː]	*vt.* 画线于……下 *n.* 下方划线
reverse	[rɪˈvɜːs]	*vi.*（使）反转；（使）颠倒 *n.* 倒转，反向
default	[dɪˈfɔːlt]	*n.* 缺省，默认
parameter	[pəˈræmɪtə]	*n.* 参数
operator	[ˈɒpəreɪtə]	*n.* 运算符
ancestor	[ˈænsestə]	*n.* 祖先；原型
disallow	[disəˈlaʊ]	*v.* 不接受，不准
strange	[streɪndʒ]	*adj.* 奇怪的，古怪的
assign	[əˈsaɪn]	*vt.* 分配，分派，选派
distinguish	[dɪˈstɪŋgwɪʃ]	*v.* 区分，辨别
expectation	[ˌekspekˈteɪʃn]	*n.* 期待，预期；期望值
container	[kənˈteɪnə]	*n.* 容器
componentwise	[kəmˈpəʊnəntwaɪz]	*n.* 分量方式，分量形式
component	[kəmˈpəʊnənt]	*n.* 部件，组件；要素，成分 *adj.* 组成的；构成的；成分的
geometrical	[ˌdʒiːəˈmetrɪkl]	*adj.* 几何的，几何学的
dimension	[daɪˈmenʃn]	*n.* 次元，度，维
meaningless	[ˈmiːnɪŋlɪs]	*adj.* 无意义的，无价值的
programmer	[ˈprəʊgræmə]	*n.* 程序设计者，程序员
undefined	[ˌʌndɪˈfaɪnd]	*adj.* 未定义的，未阐明的，未限定的

summand	[ˈsʌmænd]	*n.* 被加数
recycle	[ˌriːˈsaikl]	*v.* 再利用；再次应用；重新使用（概念、方法等）
attempt	[əˈtempt]	*vt.* 试图；尝试
incompatible	[ˌɪnkəmˈpætəbl]	*adj.* 不兼容，不相容；矛盾，不能同时成立的
syntactic	[sɪnˈtæktɪk]	*adj.* 句法的
warning	[ˈwɔːnɪŋ]	*n.* 警告
provision	[prəˈvɪʒn]	*n.* 规定，条项
scalar	[ˈskeɪlə]	*n.* 数量，标量
		adj. 数量的，标量的
matrices	[ˈmeɪtrɪsiːz]	*n.* 矩阵（matrix 的名词复数）
list	[lɪst]	*n.* 列表
position	[pəˈzɪʃn]	*vt.* 放置，安置
comment	[ˈkɒment]	*n.* 注释
anonymous	[əˈnɒnɪməs]	*adj.* 匿名的
parentheses	[pəˈrenθəsiːz]	*n.* 圆括号
mechanism	[ˈmekənɪzəm]	*n.* 机制，机能
scope	[skəʊp]	*n.* 作用域
output	[ˈaʊtpʊt]	*n. & vt.* 输出

Phrases

scripting language	脚本语言
function argument	函数参数
calling function	调用函数
side effect	副作用
variable character	可变字符
assignment operator	赋值运算符
data type	数据类型
vector space	向量空间
inner product	内积
Boolean value	布尔值，逻辑值
return value	返回值
default value	默认值，缺省值
passing variable	传递变量
non-local variable	非局部变量
global variable	全局变量

Abbreviations

VBA（Visual Basic for Applications）　　Visual Basic 的一种宏语言

Exercises

[Ex.5] Fill in the following blanks according to the text.

1. R is more than ____________. It is ____________ for doing statistics.
2. However, when ____________ or calling functions with named arguments, you must use ____________ and cannot use ____________.
3. At some time in the past R, or its ancestor S, used underscore as ____________. This meant that the ____________ of using underscores as ____________ in multi-word variable names was not only disallowed but produced ____________.
4. The first idea of a vector is what I will call ____________. This is an ordered collection of numbers ____________, such as the vector<> container in C++.
5. The R language is designed around the ____________ that a vector is an ordered set of measurements rather than ____________ or ____________.
6. The R language has no provision for ____________. The only way to represent a single number in a variable is to ____________.
7. The type of a vector is the type of ____________ it contains and must be one of the following:____________, integer, ____________, ____________, character, or ____________ All elements of a vector must have ____________.
8. Lists are created using ____________. Elements can be accessed by ____________. Named elements may be accessed either by ____________ or by ____________.
9. In a sense, R does not support ____________, only ____________. But you can change ____________of ____________, essentially making it a matrix.
10. Function arguments are passed by ____________. The most common ____________ for passing variables by reference is to ____________.

参考译文

Python 编程语言

Python 具有动态类型系统和自动内存管理功能。它支持多种编程范例，包括面向对象、命令式、函数和过程，并且具有大型且全面的标准库。

1. 特点和原理

Python 是一种多范式编程语言，完全支持面向对象的编程和结构化编程，其许多功能支持函数式编程和面向方面编程。许多其他范例通过扩展得到支持，包括契约式设计和逻辑编程。

Python 使用动态类型，并组合引用计数和循环检测垃圾收集器进行内存管理。它还具有动态名称解析（后期绑定）的特点，在程序执行期间绑定方法和变量名称。

2. 句法和语义

Python 是一种公认的易于阅读的语言，它的格式看起来整洁，在其他语言使用标点符号的地方通常使用英语关键字。与许多其他语言不同，它不使用大括号来分隔块，而语句后面的分号是可选的。它比 C 语言或 Pascal 语言具有更少的语法异常和特殊情况。

2.1 缩进

Python 使用空格缩进而不是大括号或关键字来界定块。某些语句之后会出现缩进增加的情况；缩进减少表示当前块的结束。因此，程序的视觉结构准确地表示了程序的语义结构，此功能有时也被称为缩排规则。

2.2 语句和控制流程

Python 的语句包括：

- 赋值语句（标记“=”，等号）；
- if 语句，有条件地执行代码块，也可以与 else 和 elif（else-if 的缩写）一起使用；
- for 语句，迭代可迭代对象，将每个元素捕获到局部变量以供附加块使用；
- while 语句，只要条件为真，就执行一段代码；
- raise 语句，用于引发指定的异常或重新引发捕获的异常；
- class 语句，它执行代码块并将其本地名称空间附加到类中，以用于面向对象的编程；
- def 语句，定义函数或方法；
- pass 语句，用作 NOP。在语法上需要创建一个空代码块；
- assert 语句，在调试期间用于检查应该应用的条件；
- yield 语句，它从生成器函数返回一个值。从 Python 2.5 开始，yield 也是一个运算符；
- import 语句，用于导入在当前程序中可用的函数或变量模块。有三种使用导入的方法：

import <module name> [as <alias>]或 from <module name> import *或 from <module name> import <definition 1> [as <alias 1>]，<definition 2> [as <alias 2>]，……

- print 语句已更改为 Python 3 中的 print()函数。

2.3 表达式

Python 的有些表达式与 C 和 Java 等语言类似，而有些则不同。

- 加法、减法和乘法是相同的，但除法的行为不同。Python 有两种除法类型。Python 还添加了**运算符用于取幂。

● 从 Python 3.5 开始，引入了新的@interix 运算符。它旨在把诸如 NumPy 的库用于矩阵乘法。

● 在 Python 中，==按值进行比较，而 Java 按值比较数字并按照引用比较对象。Python 的 is 运算符可用于比较对象标识。在 Python 中，可以链式比较，如<= b <= c。

● Python 在布尔运算中使用单词 and、or、not，而不使用符号&&、||、!，这与 Java 语言和 C 语言不同。

● Python 有一种称为列表推导的表达式。Python 2.4 将列表推导扩展为更通用的表达式，称为生成器表达式。

● Python 中的条件表达式写为 x if c else y （不同于许多其他语言通常使用的 c？ x：y 形式）。

● Python 区分列表和元组。列表写为[1,2,3]，是可变的，不能用作字典的键（字典键必须在 Python 中不可变）。元组写为（1,2,3），是不可变的，因此可以用作字典的键，前提是元组的所有元素都是不可变的。

● Python 有一个“字符串格式”运算符%。该功能类似于 C 中的 printf 格式字符串，例如“spam =%s eggs =%d” %（“blah”，2）。

● Python 有各种字符串文字：

① 由单引号或双引号分隔的字符串。这两种字符串都使用反斜杠（\）作为转义字符。

② 三引号字符串，以一系列三个单引号或双引号开头和结尾。

③ 原始字符串，用字符串文字前缀 r 表示。不解释转义序列。

3．库

Python 的大型标准库通常被认为是其最大的优势之一，它提供了适合许多任务的工具。对于面向互联网的应用程序，支持许多标准格式和协议，如多用途互联网邮件扩展和超文本传输协议。它包括用于创建图形用户界面、连接到关系数据库、生成伪随机数、具有任意精度小数的算术、操纵正则表达式和单元测试的模块。

标准库的某些部分包含在规范中，但大多数模块都没有。它们由代码、内部文档和测试套件（如果提供）指定。但是，由于大多数标准库是跨平台的 Python 代码，因此只有少数模块需要更改或重写。

截至 2018 年 3 月，Python Package Index（PyPI），即第三方 Python 软件的官方库，包含 130 000 多个包，具有广泛的功能，包括：

- 图形用户界面
- 网络框架
- 多媒体
- 数据库
- 联网
- 测试框架

- 自动化
- 网络抓取
- 文档建立
- 系统管理
- 科学计算
- 文本处理
- 图像处理

4. 开发环境

大多数 Python 实现（包括 CPython）都包含一个读取—求值—输出循环（REPL），允许它们作为命令行解释器运行，用户可以按顺序输入语句并立即接收结果。

其他 shell 增加了更多功能，如自动完成、会话状态保留和语法突出显示。

除标准桌面集成开发环境外，还有基于网络浏览器的 IDE：SageMath，用于开发科学和数学相关的 Python 程序；PythonAnywhere，一个基于浏览器的 IDE 和托管环境，以及 Canopy IDE，一个强调科学计算的商业 Python IDE。

Unit 6

Text A
Basic Concepts of Database

1. Data, Database and DBMS

In computer science, data is anything in a form suitable for use with a computer. Data is often distinguished from programs. A program is a set of instructions that detail a task for the computer to perform. In this sense, data is thus everything that is not program code.

A database is a collection of information that is organized so that it can easily be accessed, managed, and updated. In one view, databases can be classified according to types of content: Bibliographic, full-text, numeric, and images.

In computing, databases are sometimes classified according to their organizational approaches. The most prevalent approach is the relational database, a tabular database in which data is defined so that it can be reorganized and accessed in a number of different ways. A distributed database is one that can be dispersed or replicated among different points in a network. An object-oriented programming database is one that is congruent with the data defined in object classes and subclasses.

As one of the oldest components associated with computers, the database management system (DBMS) is a computer software program that is designed as the means of managing all databases that are currently installed on a system hard drive or network. Different types of database management systems exist, some of which are designed for the oversight and proper control of databases that are configured for specific purposes.

In database management system (DBMS), data files are the files that store the database information, whereas other files, such as index files and data dictionaries, store administrative information, known as metadata.

2. Relational Database

A relational database is a collection of data items organized as a set of formally-described tables from which data can be accessed or reassembled in many different ways without having to reorganize the database tables. The relational database was invented by E. F. Codd at IBM in 1970.

The standard user and application program interface to a relational database is the structured query language (SQL). SQL statements are used both for interactive queries for information from a relational database and for gathering data for reports.

In addition to being relatively easy to create and access, a relational database has the important advantage of being easy to extend. After the original database creation, a new data category can be added without requiring that all existing applications be modified.

A relational database is a set of tables containing data fitted into predefined categories. Each table (which is sometimes called a relation) contains one or more data categories in columns. Each row contains a unique instance of data for the categories defined by the columns. For example, a typical business order entry database would include a table that describes a customer with columns for name, address, phone number, and so forth. Another table would describe an order: Product, customer, date, sales price, and so forth. A user of the database could obtain a view of the database that meets the user's need. For example, a branch office manager might like a view or report on all customers that has bought products after a certain date. A financial services manager in the same company could, from the same tables, obtain a report on accounts that needs to be paid.

When creating a relational database, you can define the domain of possible values in a data column and further constraints that may apply to that data value. For example, a domain of possible customers could allow up to ten possible customer names but be constrained in one table to allow only three of these customer names to be specifiable.

The definition of a relational database results in a table of metadata or formal descriptions of the tables, columns, domains, and constraints.

3. Field

In database, columns of a table have special names. Each column is called a "field". Each field has a special subject. Relational databases arrange data as sets of database records. Each record consists of several fields; the fields of all records form the columns. For example, in an "address list" database, such as "Name" "Phone" "Address" "Post Code" and so on, so these columns are called as "Name" field, "Phone" field, "Address" field, "Post Code" field.

In object-oriented programming, field is also called date member, and it is the data encapsulated within a class or object. There is an instance variable for each regular field in the case. For example, a student class has a student ID field and there is one distinct ID per student. A static field is one variable, which is shared by all instances.

4. Index

We can use index to access the specific information quickly in a database table. In a database, index is a data structure, which orders data of one or more columns in a database table. For example, for name column in an employee table, if a developer wants to search specific staff with name, and compares it with table, index can help the developer access the information more quickly.

Index is a database structure. It can be created by using one or more columns of a database table, providing the basis for both rapid random search and efficient access of ordered records.

Index can provide the data pointer and then order these pointers according to the specified sort order. Index in a database is very similar to the index in using books: It searches index and finds out the special value, then finds out the column in this value.

In a relational database, an index is a copy of part of a table. In "INDEX/KEY" property page of the selected table, developers can CREATE, EDIT, or DELETE the type of each index. Indexes may be defined as unique or non-unique. A unique index acts as a constraint on the table by preventing duplicate entries in the index. When storing the index to the attach table or in relational graph where the table is, index will be stored in the database.

5. SQL

Abbreviation of structured query language. SQL is a standardized query language for requesting information from a database. The original version called SEQUEL (structured English query language) was designed by an IBM research center in 1974 and 1975. SQL was first introduced as a commercial database system in 1979 by Oracle Corporation.

Historically, SQL has been the favorite query language for database management systems running on minicomputers and mainframes. Increasingly, however, SQL is being supported by PC database systems because it supports distributed databases (databases that are spread out over several computer systems). This enables several users on a local area network to access the same database simultaneously.

Although there are different dialects of SQL, it is nevertheless the closest thing to a standard query language that currently exists. In 1986, ANSI approved a rudimentary version of SQL as the official standard, but most versions of SQL since then have included many extensions to the ANSI standard. In 1991, ANSI updated the standard.

6. Distributed Database

A distributed database is a database that is under the control of a central database management system (DBMS) in which storage devices are not all attached to a common CPU. It may be stored in multiple computers located in the same physical location, or may be dispersed over a network of interconnected computers.

Collections of data (e.g. in a database) can be distributed across multiple physical locations. A distributed database can reside on network servers on the Internet, on corporate intranets or extranets, or on other company networks. The replication and distribution of databases improve database performance at end-user worksites.

To ensure that the distributive databases are up to date and current, there are two processes: Replication and duplication. Replication involves using specialized software that looks for changes in the distributive database. Once the changes have been identified, the replication process makes all the databases look the same. The replication process can be very complex and time consuming depending on the size and number of the distributive databases. This process can also require a lot of time and computer resources. Duplication on the other hand is not as complicated. It basically identifies one database as a master and then duplicates that database. The duplication process is normally done at a set time. This is to ensure that each distributed location has the same data. In the duplication process, only changes to the master database are allowed. This is to ensure that local data will not be overwritten. Both of the processes can keep the data current in all distributive locations.

New Words

access	['ækses]	*vt.* 存取，访问
bibliographic	[ˌbɪbliə'ɒgræfɪk]	*adj.* 提要的
full-text	[fʊl'tekst]	*n.* 全文本
approach	[ə'prəʊtʃ]	*n.* 方法，步骤，途径
reorganize	[rɪ'ɔːgənaɪz]	*v.* 重新组织，再编
dispersal	[dɪ'spɜːsl]	*n.* 分布，分散
replicate	['replɪkeɪt]	*v.* 复制
congruent	['kɒŋgrʊənt]	*adj.* 适合的
subclass	['sʌbklɑːs]	*n.* 子集，子类
oversight	['əʊvəsaɪt]	*n.* 疏忽，失败
reassemble	[ˌriː ə'sembl]	*v.* 重新组织，重新装配
extend	[ɪk'stend]	*v.* 扩展
modify	['mɒdɪfaɪ]	*vt.* 更改，修改
relation	[rɪ'leɪʃn]	*n.* 关系
column	['kɒləm]	*n.* 列
row	[rəʊ]	*n.* 行
instance	['ɪnstəns]	*n.* 实例
account	[ə'kaʊnt]	*n.* 账，账目；存款
constraint	[kən'streɪnt]	*n.* 约束，强制
value	['væljuː]	*n.* 值
specifiable	['spesɪfaɪəbl]	*adj.* 可指明的，可列举的
definition	[ˌdefɪ'nɪʃn]	*n.* 定义
encapsulate	[ɪn'kæpsjʊleɪt]	*vt.* 封装

class	[klɑːs]	*n.* 类
object	[ˈɒbdʒɪkt]	*n.* 目标，对象
regular	[ˈregjʊlə]	*adj.* 有规律的；规则的，整齐的；不变的，经常使用的；固定的
distinct	[dɪˈstɪŋkt]	*adj.* 清楚的，明显的，截然不同的，独特的
specific	[spɪˈsɪfɪk]	*adj.* 具体的
employee	[ˌemplɔɪˈiː]	*n.* 雇工，雇员，职工，员工
developer	[dɪˈveləpə]	*n.* 开发人员，开发者
search	[sɜːtʃ]	*n.* 搜寻，查究 *v.* 搜索，搜寻
pointer	[ˈpɔɪntə]	*n.* 指针
define	[dɪˈfaɪn]	*vt.* 定义
prevent	[prɪˈvent]	*vt.* 防止，阻止
duplicate	[ˈdjuːplɪkɪt]	*adj.* 复制的，副的 *n.* 复制品，副本
	[ˈdjuːplɪkeɪt]	*vt.* 复制
commercial	[kəˈmɜːʃl]	*adj.* 商用的
historically	[hɪˈstɒrɪklɪ]	*adv.* 从历史角度；在历史上；以历史观点；根据历史事实
increasingly	[ɪnˈkriːsɪŋlɪ]	*adv.* 日益，愈加
dialect	[ˈdaɪəlekt]	*n.* 方言，土语；语调；[语]语支；专业用语
approve	[əˈpruːv]	*vt.* 同意，认可
approved	[əˈpruːvd]	*adj.* 经核准的，被认可的
version	[ˈvɜːʃn]	*n.* 版本
official	[əˈfɪʃl]	*adj.* 官方的，法定的
distribute	[dɪˈstrɪbjuːt]	*vt.* 分布，分发
reside	[rɪˈzaɪd]	*vi.* 住，居住，驻留，驻在
worksite	[ˈwɜːksaɪt]	*n.* 工作场所
distributive	[dɪˈstrɪbjʊtɪv]	*adj.* 分布的
replication	[ˌreplɪˈkeɪʃn]	*n.* 复制
complicated	[ˈkɒmplɪkeɪtɪd]	*adj.* 复杂的，难解的

Phrases

distinguish from	区别，识别
a suit of	一套
distributed database	分布式数据库

object-oriented programming database	面向对象编程数据库
hard drive	硬盘
index file	索引文件
data dictionary	数据字典
application program interface	应用程序接口
fit into	适合
sales price	销售价格
and so forth	等等；诸如此类；依此类推
meet one's need	满足某人的需求
branch office manager	分店业务经理
financial services manager	金融服务经理
up to	多达
instance variable	实例变量
static field	静态字段
data structure	数据结构
sort order	排序
be similar to	类似于
relational graph	关系图
spread out	扩散
under the control of	在……的控制之下，受……控制
be attached to	附加到
time consuming	耗费时间的
depending on	取决于
on the other hand	另一方面

Abbreviations

SEQUEL（Structured English QUEry Language）　结构化英语查询语言

Exercises

【Ex.1】Answer the following questions according to the text.

1. What is data in computer science?
2. What are the databases that are classified according to types of content?
3. What does DBMS stand for? What is it?
4. What is a relational database?

5. What is the standard user and application program interface to a relational database?
6. What is field in object-oriented programming?
7. What is index in database?
8. What does a unique index act as?
9. What does SQL stand for? What is it?
10. What is a distributed database?

【Ex.2】Translate the following terms or phrases from English into Chinese and vice versa.

1. data structure	1.
2. access	2.
3. distributed database	3.
4. index file	4.
5. constraint	5.
6. 面向对象编程数据库	6.
7. *n.* 关系	7.
8. *v.* 复制	8.
9. *n.* 搜寻，查究 *v.* 搜索，搜寻	9.
10. *n.* 子集，子类	10.

【Ex.3】Translate the following passages into Chinese.

The Key Elements of a Data Dictionary

A Data Dictionary provides information about each attribute, also referred to as fields, of a data model. An attribute is a place in the database that holds information. For example, if we are to create a Data Dictionary representing the articles here on Bridging the Gap, we'd potentially have attributes for article title, article author, article category, and the article content itself.

A Data Dictionary is typically organized in a spreadsheet format. Each attribute is listed as a row in the spreadsheet and each column labels an element of information that is useful to know about the attribute.

Let's look at the most common elements included in a data dictionary.

Attribute Name — A unique identifier, typically expressed in business language, that labels each attribute.

Optional/Required — Indicates whether information is required in an attribute before a record can be saved.

Attribute Type — Defines what type of data is allowable in a field. Common types include text, numeric, date/time, enumerated list, look-ups, booleans, and unique identifiers.

While these are the core elements of a data dictionary, it's not uncommon to document additional information about each element, which may include the source of the information, the table or concept in which the attribute is contained, the physical database field name, the field length, and any default values.

【Ex.4】Fill in the blanks with the words given below.

quantification	find	applied	analysis	Intuitively
communicating	as	quantum	expressed	than

Information theory is a branch of ___1___ mathematics and engineering involving the ___2___ of information. Historically, information theory was developed to find fundamental limits on compressing and reliably ___3___ data. Since its inception it has broadened to ___4___ applications in many other areas, including statistical inference, networks other ___5___ communication networks as in neurobiology, the evolution and function of molecular codes, model selection in ecology, thermal physics, ___6___ computing, plagiarism detection and other forms of data ___7___.

A key measure of information that comes up in the theory is known ___8___ information entropy, which is usually ___9___ by the average number of bits needed for storage or communication. ___10___, entropy quantifies the uncertainty involved in a random variable. For example, a fair coin flip will have less entropy than a roll of a die.

Text B
Database Management System

扫码听课文

1. Definition

A database management system (DBMS) is the software that allows a computer to perform database functions of storing, retrieving, adding, deleting and modifying data. Relational database management systems (RDBMS) implement the relational model of tables and relationships.

2. Types of DBMS

There are four main types of database management systems (DBMS) and these are based upon their management of database structures. In other words, the types of DBMS are entirely dependent upon how the database is structured by that particular DBMS.

2.1 Hierarchical DBMS

A DBMS is said to be hierarchical if the relationships among data in the database are established in such a way that one data item is present as the subordinate of another one or a sub unit. Here subordinate means that items have "parent-child" relationships among them. Direct relationships exist between any two records that are stored consecutively. The data structure "tree" is followed by the DBMS to structure the database. No backward movement is possible or allowed in the hierarchical database.

The hierarchical data model was developed by IBM in 1968 and introduced into information management systems. This model is like a structure of a tree with the records forming the nodes and fields forming the branches of the tree. In the hierarchical model, records are linked in the form of an organization chart. A tree structure may establish one-to-many relationship.

2.2 Network DBMS

A DBMS is said to be a Network DBMS if the relationships among data in the database are of type many-to-many. The relationships among many-to-many appear in the form of a network. Thus the structure of a network database is extremely complicated because of these many-to-many relationships in which one record can be used as a key of the entire database. A network database is structured in the form of a graph that is also a data structure. Though the structure of such a DBMS is highly complicated, it has two basic elements i.e. records and sets to designate many-to-many relationships. Mainly high-level languages such as Pascal, COBOL and FORTRAN, etc., are used to implement the records and set structures.

2.3 Relational DBMS

RDBMS, short for relational database management system and pronounced as separate letters, is a type of database management system that stores data in the form of related tables. It is a program that lets you create, update, and administer a relational database.

Relational databases are powerful because they require few assumptions about how data is related or how it will be extracted from the database. As a result, the same database can be viewed in many different ways.

An important feature of relational systems is that a single database can be spread across several tables. This differs from flat-file databases in which each database is self-contained in a single table.

Almost all full-scale database systems are RDBMSs. Small database systems, however, use other designs that provide less flexibility in posing queries. A number of RDBMSs are available. Some popular examples are Oracle, Sybase, Ingress, Informix, Microsoft SQL Server, and Microsoft Access.

2.4 Object-Oriented DBMS

An object-oriented DBMS is able to handle many new data types, including graphics, photographs, audios, and videos. Object-oriented databases represent a significant advance over their other database cousins. Hierarchical and network databases are all designed to handle structured data; that is, data that fits nicely into fields, rows, and columns. They are useful for handling small snippets of information such as names, addresses, zip codes, product numbers, and any kind of statistics or numbers you can think of. On the other hand, an object-oriented database can be used to store data from a variety of media sources, such as photographs and text, and produce work, as output, in a multimedia format.

3. Advantages of Database Management System

3.1 Data Independence

Application programs should be as independent as possible from details of data representation and storage. The DBMS can provide an abstract view of the data to insulate application code from such details.

3.2 Efficient Data Access

A DBMS utilizes a variety of sophisticated techniques to store and retrieve data efficiently. This feature is especially important if the data is stored on external storage devices.

3.3 Data Integrity and Security

If data is always accessed through the DBMS, the DBMS can enforce integrity constraints on the data. For example, before inserting salary information for an employee, the DBMS can check that the department budget is not exceeded. Also, the DBMS can enforce access controls that govern what data is visible to different classes of users.

3.4 Data Administration

When several users share the data, centralizing the administration of data can offer significant improvements. Experienced professionals, who understand the nature of the data being managed and how different groups of users use it, can be responsible for organizing the data representation to minimize redundancy and for retuning the storage of the data to make retrieval efficient.

3.5 Concurrent Access and Crash Recovery

A DBMS schedules concurrent accesses to the data in such a manner that users can think of the data as being accessed by only one user at a time. Further, the DBMS protects users from the effects of system failures.

3.6 Reduced Application Development Time

Clearly, the DBMS supports many important functions that are common to many applications accessing data stored in the DBMS. This, in conjunction with the high-level interface to the data, facilitates quick development of applications. Such applications are also likely to be more robust than applications developed from scratch because many important tasks are handled by the DBMS instead of being implemented by the application.

4. DBMS Functions

There are several functions that a DBMS performs to ensure data integrity and consistency in the database. The ten functions in the DBMS are: Data dictionary management, data storage management, data transformation and presentation, security management, multiuser access control, backup and recovery management, data integrity management, database access languages and application programming interfaces, database communication interfaces, and transaction management.

4.1 Data Dictionary Management

Data Dictionary is where the DBMS stores definitions of the data elements and their relationships (metadata). The DBMS uses this function to look up the required data component structures and relationships. When programs access data in a database they are basically going through the DBMS. This function removes structural and data dependency and provides the user with data abstraction. In turn, this makes things a lot easier for the end user. The Data Dictionary is often hidden from the user and is used by Database Administrators and Programmers.

4.2 Data Storage Management

This particular function is used for the storage of data and any related data entry forms or screen definitions, report definitions, data validation rules, procedural code, and structures that can handle video and picture formats. Users do not need to know how data is stored or manipulated. Also involved with this structure is a term called performance tuning that relates to a database's efficiency in relation to storage and access speed.

4.3 Data Transformation and Presentation

This function exists to transform any data entered into required data structures. By using the data transformation and presentation function the DBMS can determine the difference between logical and physical data formats.

4.4 Security Management

This is one of the most important functions in the DBMS. Security management sets rules that determine specific users that are allowed to access the database. Users are given a username and password or sometimes through biometric authentication (such as a fingerprint or retina scan), but these types of authentication tend to be more costly. This function also sets restraints on what specific data any user can see or manage.

4.5 Multi-User Access Control

Data integrity and data consistency are the basis of this function. Multi-user access control is a very useful tool in a DBMS, it enables multiple users to access the database simultaneously without affecting the integrity of the database.

4.6 Backup and Recovery Management

Backup and recovery is brought to mind whenever there is potential outside threats to a database. For example if there is a power outage, recovery management is how long it takes to recover the database after the outage. Backup management refers to the data safety and integrity; for example, backing up all your mp3 files on a disk.

4.7 Data Integrity Management

The DBMS enforces these rules to reduce things such as data redundancy, which is when data is stored in more than one place unnecessarily, and maximize data consistency, make sure database is returning correct/same answer each time for same question asked.

4.8 Database Access Languages and Application Programming Interfaces

A query language is a nonprocedural language. An example of this is SQL (structured query language). SQL is the most common query language supported by the majority of DBMS vendors. The use of this language makes it easy for users to specify what they want done without the headache of explaining how to specifically do it.

4.9 Database Communication Interfaces

This refers to how a DBMS can accept different end user requests through different network environments. An example of this can be easily related to the Internet. A DBMS can provide access to the database using the Internet through Web Browsers (Mozilla Firefox, Internet Explorer, and Netscape).

4.10 Transaction Management

This refers to how a DBMS must supply a method that will guarantee that all the updates in a given transaction are made or not made. All transactions must follow what is called the ACID properties.

A — Atomicity: A transaction is an indivisible unit that is either performed as a whole and not by its parts, or not performed at all. It is the responsibility of recovery management to make sure this takes place.

C — Consistency: A transaction must alter the database from one constant state to another constant state.

I — Isolation: Transactions must be executed independently of one another. Part of a transaction in progress should not be able to be seen by another transaction.

D — Durability: A successfully completed transaction is recorded permanently in the database and must not be lost due to failures.

New Words

subordinate	[sə'bɔːdɪnət]	*adj.* 次要的，从属的，下级的
consecutively	[kən'sekjʊtɪvlɪ]	*adv.* 连续地
record	['rekɔːd]	*n.* 记录
branch	[brɑːntʃ]	*n.* 枝，分枝
designate	['dezɪgneɪt]	*v.* 指定，指派
self-contained	[self-kən'teɪnd]	*adj.* 设备齐全的，独立的；自给自足的
graphics	['græfɪks]	*n.* 图形
photograph	['fəʊtəgrɑːf]	*n.* 照片
audio	['ɔːdɪəʊ]	*adj.* 音频的，声频的
video	['vɪdɪəʊ]	*n.* 视频
extremely	[ɪk'striːmlɪ]	*adv.* 极端地，非常地
snippet	['snɪpɪt]	*n.* 小片，片段
statistic	[stə'tɪstɪk]	*n.* 统计量 *adj.* 统计的，统计学的
independence	[ˌɪndɪ'pendəns]	*n.* 独立性
abstract	['æbstrækt]	*adj.* 抽象的 *vt.* 提炼，抽象化
insulate	['ɪnsjʊleɪt]	*vt.* 隔离
centralize	['sentrəlaɪz]	*vt.* 集聚，集中
improvement	[ɪm'pruːvmənt]	*n.* 改进，进步
retune	['riː'tjuːn]	*v.* 重新调整，重新调节
concurrent	[kən'kʌrənt]	*adj.* 并发的，协作的，一致的
crash	[kræʃ]	*v.* 崩溃，垮台
recovery	[rɪ'kʌvərɪ]	*n.* 恢复，防御
multiuser	[ˌmʌltɪ'juːzə]	*n.* 多用户
remove	[rɪ'muːv]	*vt.* 删除，移动
biometric	[baɪəʊ'metrɪk]	*adj.* 生物特征识别
fingerprint	['fɪŋgəprɪnt]	*n.* 指纹，手印 *vt.* 采指纹
restraint	[rɪ'streɪnt]	*n.* 抑制，制止，克制
affect	[ə'fekt]	*vt.* 影响
outage	['aʊtɪdʒ]	*n.* 中断，断供
unnecessarily	[ʌn'nesəsərəlɪ]	*adv.* 不必要地，未必
majority	[mə'dʒɒrɪtɪ]	*n.* 多数，大半

guarantee	[ˌgærən'tiː]	*n.* & *vt.* 保证
atomicity	[ˌætə'mɪsɪtɪ]	*n.* 原子性
indivisible	[ˌɪndɪ'vɪzəbl]	*adj.* 不能分割的，除不尽的
consistency	[kən'sɪstənsɪ]	*n.* 坚固性，一致性

Phrases

relational model	关系模型
base upon	根据，依据
be established in	立足于
data item	数据项
hierarchical data model	层次数据模型
organization chart	组织系统图，编组表
one-to-many relationship	一对多关系
many-to-many relationship	多对多关系
high-level language	高级语言
object-oriented database	面向对象数据库
a variety of	多种的
sophisticated technique	现有技术，成熟技术
external storage device	外部存储设备
minimize redundancy	最小化冗余
concurrent access	并行访问
protect from	保护
system failure	系统故障，系统损坏
in conjunction with …	与……协力
from scratch	从零开始，从无到有
data transformation	数据转换
data element	数据元素
data validation rule	数据验证规则
enter into	进入，参加，成为……一部分
physical data	物理数据
Multi-User Access Control	多用户访问控制
power outage	停电
back up	备份
make sure	确保，确定
query language	查询语言
communication interface	通信接口

end user　终端用户，最终用户
independent of …　不依赖……，独立于……
in progress　进行中

Abbreviations

RDBMS（Relational DataBase Management System）　关系数据库管理系统
ACID（Atomicity, Consistency, Isolation, Durability）　原子性、一致性、隔离性、持久性

Exercises

【Ex.5】Answer the following questions according to the text.

1. What does RDBMS stand for? What does it do?
2. How many main types of database management systems (DBMS) are there? What are they based upon?
3. Who developed the hierarchical data model? When?
4. Why is the structure of a network database extremely complicated?
5. Why are relational databases powerful?
6. What can an object-oriented database be used to do?
7. What are the advantages of a database management system?
8. How many functions does a DBMS perform to ensure data integrity and consistency of data in the database? What are they?
9. What does security management do?
10. What does transaction management refer to? What must all transactions follow?

参考译文

数据库基本概念

1. 数据、数据库和数据库管理系统

在计算机科学中，适合计算机使用的任何形式的东西都是数据。数据通常与程序不同。程序是一组指令，详细地描述了计算机要执行的任务。在这个意义上说，不是程序代码的东西都是数据。

数据库是所组织的信息的集合，这样可以很容易地访问、管理和更新这些信息。有一种观点认为，数据库可以根据其内容分为以下几类：概要、文字、数字和图像。

在计算中，有时也根据数据库的组织方法对其分类。最普遍的组织方法是关系数据库——

表式数据库，在这个数据库中定义数据，可以用不同的方式进行重组和访问。分布式数据库分散在网络中的不同位置，可以有多个副本。一个面向对象编程数据库与对象类和子类中定义的数据相一致。

作为与计算机相关的最古老组件之一，数据库管理系统（DBMS）是一个计算机软件程序，用来管理安装在系统硬盘驱动器或网络中的所有数据库。目前有不同类型的数据库管理系统，其中一些用来对为特定目的而配置的数据库进行监管和适当的控制。

在数据库管理系统中，数据文件是存储数据库信息的文件，而其他文件（如索引文件和数据字典）存储管理信息，称为元数据。

2. 关系数据库

关系数据库是数据项的集合，把这些数据项组织成一组形式化描述表格，从中可以用多种方法访问和重新组合这些数据而无需重新组织数据库表。IBM 的 EF Codd 于 1970 年创造了关系数据库。

关系数据库的标准用户和应用程序接口是结构化查询语言（SQL）。SQL 语句既用于关系数据库信息的交互查询，也用于收集报表数据。

除了相对容易创建和访问外，关系数据库的另一重要优势是易于扩展。在原始数据库创建后，可以增加新的数据类型，而无须对现有的全部应用程序进行修改。

关系数据库是一组数据表，这些数据表包含了符合预定义类别的数据。每个表的（有时也被称为关系）列包含一类或多类数据，每一行都包含唯一的按列分类的数据实例。例如，一个典型的业务订单登记数据库包括一个表，按列描述了一个顾客姓名、地址、电话号码等信息。另一张表描述的顺序为：产品、客户、日期、销售价格等。数据库用户可以按照自己的需要查看数据库。例如，一个分公司经理可能要查看某一特定日期后购买了产品的所有客户，或形成报表。同一公司一个财务经理从同一个表可以得到需要支付的账户报表。

当创建一个关系数据库时，可以定义数据列的取值范围，并进一步限制可用的数据值。例如，客户域最多可以有 10 个客户名称，但限制在一个表中只能指定三个客户名称。

定义一个关系数据库会产生一个元数据表，或者说是形式化地描述表、列、域和约束。

3. 字段

在数据库中，表的列有特殊的名字。每一列被称为一个“字段”。每个字段都有一个特别的主题。关系数据库按照数据库记录集排列数据。每个记录包含几个字段，所有记录的字段形成列。例如，在“地址清单”的数据库中，如“姓名”“电话”“地址”“邮政编码”等，所以这些列被称为“姓名”字段、“电话”字段、“地址”字段、“邮政编码”字段。

在面向对象的编程中，字段也称为数据成员，并且被封装在一个类别或对象中。在这种情况下，每个正则字段都有一个实例变量。例如，一个学生类有一个学生 ID 字段，每个学生有一个独特的 ID。静态字段是一个变量，它为所有实例所共享。

4. 索引

我们可以使用索引迅速访问数据库表中的特定信息。在数据库中，索引是一种数据结构，对数据库表中一列或多列数据排序。例如，雇员表中的名称栏，如果开发人员希望搜索特定人员的名称，并与表相比较，索引可以帮助开发人员更快速地访问相关信息。

索引是一种数据库结构，可以使用一个或多个数据库表列来创建它，它是快速随机查看和对有序记录进行高效访问的基础。

索引可以提供数据指针，然后根据指定的顺序对其排序。在一个数据库中使用索引与在书籍中使用索引颇为相似：它检索索引并找出特殊值，然后找出该值的列。

在关系数据库中，索引是表的部分副本。在选定表的“INDEX/KEY”属性页，开发人员可以 CREATE（创建）、EDIT（编辑）或 DELETE（删除）每个索引的类型，可以唯一地或者非唯一地定义索引。唯一索引通过防止索引中出现重复项对表进行限定。当把索引存储到关联表或存储到表的关系图时，索引将被存储在数据库中。

5. SQL

SQL 是“结构化查询语言”的缩写。SQL 是一种从数据库查询信息的标准化查询语言。原来的版本称为 SEQUEL（结构化英语查询语言），是由 IBM 研究中心在 1974 年和 1975 年设计的。SQL 作为商业数据库系统于 1979 年首次由甲骨文公司推出。

从历史上看，SQL 一直是运行在小型机和大型机上数据库管理系统所喜爱的查询语言。然而，越来越多的 PC 数据库系统都支持 SQL，因为它支持分布式数据库（分布在多个计算机系统上的数据库）。这使一个局域网上的多个用户可以同时访问同一个数据库。

虽然 SQL 有多种不同的方言，但它仍然是目前与标准查询语言最接近的查询语言。1986 年，ANSI 批准了官方标准的一个基本版本的 SQL，但从此之后，大多数 SQL 版本都对那时 ANSI 的标准做了扩展。1991 年，ANSI 更新了标准。

6. 分布式数据库

分布式数据库是由中央数据库管理系统（DBMS）控制的数据库，在该数据库中，存储设备并不都连接到一个共同的 CPU。数据库可以存储在同一物理位置的多台计算机上，也可以存储在通过网络互连的分散计算机上。

数据集合（例如，在数据库中）可以分布在多个物理位置。一个分布式数据库可以驻留在互联网、企业内联网或外联网的网络服务器上。数据库的复制和分散提高了最终用户工作场所的数据库性能。

要确保分布数据库的及时更新和即时性，有两个过程：复制和制作副本。复制涉及使用专门的软件，查找分布式数据库的变更。一旦变更确定下来，复制过程让所有的数据库看起来一样。根据分布数据库的大小和数量，复制过程可能非常复杂和耗时。这个过程还涉及大量的时间和计算机资源。而制作副本并不那么复杂，它基本上是确定主数据库，然后复制该数据库。制作副本过程通常是在设定的时间进行，这是为了确保每个分布位置具有相同的数据。在制作副本过程中，只允许主数据库的变化，这是为了保证本地数据不会被覆盖。这两个过程都可以保持各个分布位置数据的即时性。

Unit 7

Text A
Data Warehouse

In computing, a data warehouse (DW), also known as an enterprise data warehouse (EDW), is a system used for reporting and data analysis, and is considered a core component of business intelligence. DWs are central repositories of integrated data from one or more disparate sources. They store current and historical data that are used for creating analytical reports for workers throughout the enterprise in one single place.

The data stored in the warehouse are uploaded from the operational systems (such as marketing or sales). The data may pass through an operational data store and may require data cleansing for additional operations to ensure data quality before they are used in the DW for reporting (See Figure 7-1).

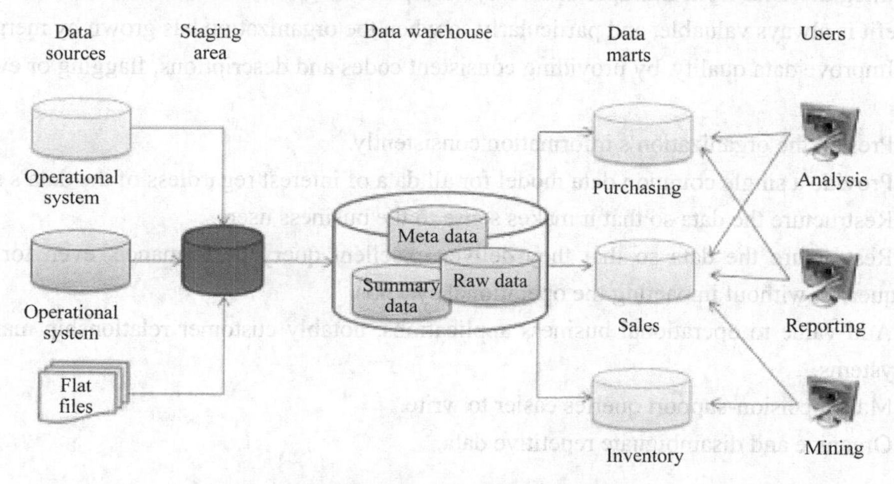

Figure 7-1 The Basic Architecture of a Data Warehouse

The typical extract, transform, load (ETL)-based data warehouse uses staging, data integration, and access layers to house its key functions. The staging layer or staging database stores raw data extracted from each of the disparate source data systems. The integration layer integrates the disparate data sets by transforming the data from the staging layer and it stores this transformed data in an operational data store (ODS) database. The integrated data are then moved to yet another database, often called the data warehouse database. The access layer helps users retrieve data.

Data are cleansed, transformed, catalogued, and made available for use by managers and other business professionals for data mining, online analytical processing, market research and decision support. However, the means to retrieve and analyze data, to extract, transform, and load data, and to manage the data dictionary are also considered essential components of a data warehousing system. Many references to data warehousing use this broader context. Thus, an expanded definition for data warehousing includes business intelligence tools, tools to extract, transform, and load data into the repository, and tools to manage and retrieve metadata.

1. Benefits

A data warehouse maintains a copy of information from the source transaction systems. This architectural complexity provides the opportunity to:

- Integrate data from multiple sources into a single database and data model, then provide more data to a single database so a single query engine can be used to present data in an ODS.
- Mitigate the problem of database isolation level lock contention in transaction processing systems caused by attempts to run large, long-running, analysis queries in transaction processing databases.
- Maintain data history, even if the source transaction systems do not.
- Integrate data from multiple source systems, enabling a central view across the enterprise. This benefit is always valuable, and particularly so when the organization has grown by merger.
- Improve data quality, by providing consistent codes and descriptions, flagging or even fixing bad data.
- Present the organization's information consistently.
- Provide a single common data model for all data of interest regardless of the data's source.
- Restructure the data so that it makes sense to the business users.
- Restructure the data so that they deliver excellent query performance, even for complex analytic queries, without impacting the operational systems.
- Add value to operational business applications, notably customer relationship management (CRM) systems.
- Make decision-support queries easier to write.
- Organize and disambiguate repetitive data.

2. Generic Environment

The environment for data warehouses and marts includes the following:

- Source systems that provide data to the warehouse or mart;

- Data integration technology and processes that are needed to prepare the data for use;
- Different architectures for storing data in an organization's data warehouse or data marts;
- Different tools and applications for the variety of users;
- Metadata, data quality, and governance processes must be in place to ensure that the warehouse or mart meets its purposes.

With regards to the source systems listed above, R. Kelly Rainer states, "A common source for the data in data warehouses is the company's operational databases, which can be relational databases".

Regarding data integration, Rainer states, "It is necessary to extract data from source systems, transform them, and load them into a data mart or warehouse".

Today, the most successful companies are those that can respond quickly and flexibly to market changes and opportunities. A key to this response is the effective and efficient use of data and information by analysts and managers. A "data warehouse" is a repository of historical data that are organized by subject to support decision makers in the organization. Once data are stored in a data mart or warehouse, they can be accessed.

3. Related Systems (Data Mart, OLAP, OLTP, Predictive Analytics)

A data mart is a simple form of a data warehouse that is focused on a single subject (or functional area), hence people draw data from a limited number of sources such as sales, finance or marketing. Data marts are often built and controlled by a single department within an organization. The sources could be internal operational systems, a central data warehouse, or external data. Denormalization is the norm for data modeling techniques in this system. Given that data marts generally cover only a subset of the data contained in a data warehouse, they are often easier and faster to implement.

Difference Between Data Warehouse and Data Mart

Attribute	Data Warehouse	Data Mart
Scope of the data	enterprise-wide	department-wide
Number of subject areas	multiple	single
How difficult to build	difficult	easy
How long it takes to build	more	less
Amount of memory	larger	limited

Types of data marts include dependent, independent, and hybrid data marts.

Online analytical processing (OLAP) is characterized by a relatively low volume of transactions. Queries are often very complex and involve aggregations. For OLAP systems, response time is an effectiveness measure. OLAP applications are widely used by Data Mining techniques. OLAP databases store aggregated, historical data in multidimensional schemas (usually star schemas). OLAP systems typically have data latency of a few hours, as opposed to data marts, where latency is expected to be closer to one day. The OLAP approach is used to analyze multidimensional data from multiple sources and perspectives. The three basic operations in OLAP are: Roll-up (Consolidation), Drill-down and Slicing & Dicing.

Online transaction processing (OLTP) is characterized by a large number of short on-line transactions (INSERT, UPDATE, DELETE). OLTP systems emphasize very fast query processing and maintaining data integrity in multi-access environments. For OLTP systems, effectiveness is measured by the number of transactions per second. OLTP databases contain detailed and current data. The schema used to store transactional databases is the entity model (usually 3NF). Normalization is the norm for data modeling techniques in this system.

Predictive analytics is about finding and quantifying hidden patterns in the data using complex mathematical models that can be used to predict future outcomes. Predictive analysis is different from OLAP in that OLAP focuses on historical data analysis and is reactive in nature, while predictive analysis focuses on the future. These systems are also used for customer relationship management (CRM).

4. Design Methods

4.1 Bottom-Up Design

In the bottom-up approach, data marts are first created to provide reporting and analytical capabilities for specific business processes. These data marts can then be integrated to create a comprehensive data warehouse.

4.2 Top-Down Design

The top-down approach is designed using a normalized enterprise data model. "Atomic" data, that is, data at the greatest level of detail, are stored in the data warehouse. Dimensional data marts containing data needed for specific business processes or specific departments are created from the data warehouse.

4.3 Hybrid Design

Data warehouses (DW) often resemble the hub and spokes architecture. Legacy systems feeding the warehouse often include customer relationship management and enterprise resource planning, and generate large amounts of data. To consolidate these various data models, and facilitate the extract transform load process, data warehouses often make use of an operational data store, the information from which is parsed into the actual DW. To reduce data redundancy, larger systems often store the data in a normalized way. Data marts for specific reports can then be built on top of the data warehouse.

5. Data Warehouse Characteristics

There are basic features that define the data in the data warehouse that include subject orientation, data integration, time-variant, nonvolatile data, and data granularity.

5.1 Subject-Oriented

Unlike the operational systems, the data in the data warehouse revolves around subjects of the enterprise (database normalization). Subject orientation can be really useful for decision making. Gathering the required objects is called subject oriented.

5.2 Integrated

The data found within the data warehouse are integrated. Since they come from several operational systems, all inconsistencies must be removed. Consistencies include naming conventions, measurement of variables, encoding structures, physical attributes of data, and so forth.

5.3 Time-Variant

While operational systems reflect current values as they support day-to-day operations, in the data warehouse, data represents data over a long time horizon (up to 10 years) which means it stores historical data. It is mainly meant for data mining and forecasting. If a user is searching for a buying pattern of a specific customer, the user needs to look at data on the current and past purchases.

5.4 Nonvolatile

The data in the data warehouse are read-only which means they cannot be updated, created, or deleted.

5.5 Summary

In the data warehouse, data are summarized at different levels.The user may start looking at the total sale units of a product in an entire region. Then the user looks at the states in that region. Finally, they may examine the individual stores in a certain state. Therefore, typically, the analysis starts at a higher level and moves down to lower levels of details.

New Words

warehouse	['weəhaʊs]	*n.* 仓库 *vt.* 把……放入或存入仓库
reporting	[rɪ'pɔːtɪŋ]	*n.* 报告 *adj.* 报告的
integrate	['ɪntɪgreɪt]	*v.* 集成，合并；成为一体
disparate	['dɪspərɪt]	*adj.* 完全不同的

source	[sɔːs]	*n.* 根源，本源；源头
		vt.（从……）获得；寻求（尤指供货）的来源
		vi. 起源；寻求来源；寻求生产商（或提供商）
upload	[ˌʌpˈləʊd]	*vt.* 上传，上载
operational	[ˌɒpəˈreɪʃənl]	*adj.* 操作的；经营的
staging	[ˈsteɪdʒɪŋ]	*n.* 临时存储
catalogue	[ˈkætəlɒg]	*vt.* 为……编目录；登记分类
essential	[ɪˈsenʃl]	*adj.* 基本的；必要的；本质的
		n. 必需品；基本要素
architectural	[ˌɑːkɪˈtektʃərəl]	*adj.* 建筑的；结构的
opportunity	[ˌɒpəˈtjuːnɪtɪ]	*n.* 机会；有利的环境，条件
mitigate	[ˈmɪtɪgeɪt]	*vt.* 使缓和，使减轻；使平息
		vi. 减轻，缓和下来
isolation	[ˌaɪsəˈleɪʃn]	*n.* 隔离；孤立状态
description	[dɪˈskrɪpʃn]	*n.* 描述；形容
fix	[fɪks]	*v.* 修理
consistently	[kənˈsɪstəntlɪ]	*adv.* 一贯地，坚持地
restructure	[ˌriːˈstrʌktʃə]	*v.* 重建；重组；调整
disambiguate	[ˌdɪsæmˈbɪgjʊeɪt]	*vt.* 消除……的歧义
repetitive	[rɪˈpetɪtɪv]	*adj.* 重复的
prepare	[prɪˈpeə]	*vt.* 准备
respond	[rɪˈspɒnd]	*v.* 回答，响应
flexibly	[ˈfleksɪblɪ]	*adv.* 灵活地
effective	[ɪˈfektɪv]	*adj.* 有效的；起作用的
efficient	[ɪˈfɪʃnt]	*adj.* 有效率的，高效的
denormalization	[ˈdiːnɔːməlaɪzeɪʃən]	*n.* 反规范化，反向规格化；非规范化
subset	[ˈsʌbset]	*n.* 子集
attribut	[ˈætrɪbjuːt]	*n.* 属性，性质，特征
	[əˈtrɪbjuːt]	*vt.* 把……归于
dependent	[dɪˈpendənt]	*adj.* 依靠的，依赖的
aggregation	[ˌægrɪˈgeɪʃn]	*n.* 聚集；集成；集结
effectiveness	[ɪˌfekˈtɪvnɪs]	*n.* 有效性；效益；效用
measure	[ˈmeʒə]	*n.* 度量，测量，测度
		vt. 测量，估量
multidimensional	[ˌmʌltɪdɪˈmenʃənl]	*adj.* 多维的；多重的
consolidation	[kənˌsɒlɪˈdeɪʃən]	*n.* 合并；联合
slicing	[ˈslaɪsɪŋ]	*n.* 切割

dicing	['daɪsɪŋ]	*n.* 切片
integrity	[ɪn'tegrɪtɪ]	*n.* 完整性
normalization	[ˌnɔːməlaɪ'zeɪʃn]	*n.* 规范化；标准化；正态化
hidden	['hɪdn]	*adj.* 隐藏的
pattern	['pætn]	*n.* 模式
bottom-up	['bɒtəm'ʌp]	*adj.* 自底向上的
top-down	[tɒp daʊn]	*adj.* 自顶向下的
atomic	[ə'tɒmɪk]	*adj.* 原子的；极微的
department	[dɪ'pɑːtmənt]	*n.* 部门
hybrid	['haɪbrɪd]	*n.* 混合物
		adj. 混合的
resemble	[rɪ'zembl]	*vt.* 与……相像，类似于
consolidate	[kən'sɒlɪdeɪt]	*vt.* 把……合成一体，合并
		vi. 统一；合并；联合
facilitate	[fə'sɪlɪteɪt]	*vt.* 促进；使容易
time-variant	['taɪm'veərɪənt]	*adj.* 时变的
nonvolatile	['nɒn'vɒlətaɪl]	*adj.* 非易失性
granularity	[grænjʊ'lærɪtɪ]	*n.* 间隔尺寸；粒度
subject-oriented	['sʌbdʒɪkt 'ɔːrɪəntɪd]	*adj.* 面向主题的
revolve	[rɪ'vɒlv]	*vt.* 反复考虑；使循环
		vi. 循环出现；反复考虑
convention	[kən'venʃn]	*n.* 惯例，习俗，规矩
measurement	['meʒəmənt]	*n.* 量度
reflect	[rɪ'flekt]	*v.* 反映，反射
day-to-day	['deɪtə'deɪ]	*adj.* 日常的，逐日的；经常
horizon	[hə'raɪzn]	*n.* 范围；界限
purchase	['pɜːtʃəs]	*v.* 购买，采购
		n. 购买；购买行为
read-only	['riːd'əʊnlɪ]	*n.* 只读
update	[ˌʌp'deɪt]	*vt.* 更新；校正，修正
delete	[dɪ'liːt]	*v.* 删除
region	['riːdʒən]	*n.* 范围，领域

Phrases

data analysis	数据分析

business intelligence	商业智能
operational system	操作系统；业务系统；运行系统；运营系统；营运系统
data integration	数据整合，数据集成
raw data	原始数据
online analytical processing	在线分析处理
decision support	决策支持
data dictionary	数据字典
business intelligence tool	商业智能工具
transaction system	事务系统，业务处理系统
query engine	查询引擎
data mart	数据集市
decision maker	决策者
response time	响应时间
data latency	数据延迟
data integrity	数据完整性
modeling technique	建模技术
predictive analytic	预测分析
mathematical model	数学模型
design method	设计方法
business process	业务流程
hub and spokes architecture	中心轮辐式架构
enterprise resource planning	企业资源规划
be parsed into	被解析为
encoding structure	编码结构
physical attribute	物理属性
buying pattern	购买模式

Abbreviations

DW（Data Warehouse）	数据仓库
EDW（Enterprise Data Warehouse）	企业数据仓库
ODS（Operational Data Store）	操作数据存储
CRM（Customer Relationship Management）	客户关系管理
OLAP（Online Analytical Processing）	联机分析处理
OLTP（Online Transaction Processing）	联机事务处理

Exercises

【Ex.1】Answer the following questions according to the text.

1. What is a data warehouse (DW) in computing?
2. What are also considered essential components of a data warehousing system?
3. What does an expanded definition for data warehousing include?
4. What does integrating data from multiple source systems enable? For whom is this benefit particularly valuable?
5. What must be in place to ensure that the warehouse or mart meets its purposes?
6. What are the most successful companies today? What is the key to this response?
7. What is a data mart? How are data marts often built and controlled?
8. What is the OLAP approach used to do? What are the three basic operations in OLAP?
9. What are the three design methods of data marts?
10. What are the data warehouse characteristics mentioned in the passage?

【Ex.2】Translate the following terms or phrases from English into Chinese and vice versa.

1. business intelligence	1. ______
2. data integration	2. ______
3. transaction system	3. ______
4. data integrity	4. ______
5. buying pattern	5. ______
6. 原始数据	6. ______
7. *n.*量度	7. ______
8. *adj.*非易失性	8. ______
9. *n.*仓库	9. ______
10. *n.*规范化；标准化；正态化	10. ______

【Ex.3】Translate the following passages into Chinese.

Data Integrity

Data integrity is the assurance that digital information is uncorrupted and can only be accessed or modified by those authorized. Integrity involves maintaining the consistency, accuracy and trustworthiness of data over its entire life cycle.

To maintain integrity, data must not be changed in transit and steps must be taken to ensure that data cannot be altered by an unauthorized person or program. Such measures include implementing user access controls and version control to prevent erroneous changes or accidental deletion by authorized users. Other measures include the use of checksums and cryptographic checksums to verify integrity. Network administration measures to ensure data integrity include documenting system administration procedures, parameters and maintenance activities, and creating disaster recovery plans for occurrences such as power outages, server failure or security attacks. Should data become corrupted, backups or redundancies must be available to restore the affected data to its correct state.

Measures must also be taken to ensure integrity by controlling the physical environment of networked terminals and servers because data consistency, accuracy and trustworthiness can also be threatened by environmental hazards such as heat, dust or electrical problems. Some means must be in place to detect any changes in data that might occur as a result of non-human-caused events such as an electromagnetic pulse or server crash. Practices followed to protect data integrity in the physical environment include keeping transmission media (such as cables and connectors) covered and protected to ensure that they cannot be tapped, and protecting hardware and storage media from power surges, electrostatic discharges and magnetism.

【Ex.4】Fill in the blanks with the words given below.

cost	online	important	removed	investment
indexed	automatically	products	volume	archive

Data Archiving

Data archiving is the process of moving data that is no longer actively used to a separate storage device for long-term retention. Archive data consist of older data that are still ____1____ to the organization and may be needed for future reference, as well as data that must be retained for regulatory compliance. Data archives are ____2____ and have search capabilities so files and parts of files can be easily located and retrieved.

The greatest benefit of archiving data is that it reduces the ____3____ of primary storage. Primary storage is typically expensive because a storage array must produce a sufficient level of IOPS to meet operational requirements for user read/write activity. In contrast, ____4____ storage costs less because it is typically based on a low-performance, high-capacity storage medium.

Archive storage also reduces the ____5____ of data that must be backed up. Removing infrequently accessed data from the backup data set improves backup and restore performance, and lowers secondary storage costs.

Data archives take a number of different forms. Some systems make use of ____6____ data storage, which places archive data onto disk systems where they are readily accessible. Archives are frequently file-based, but object storage is growing in popularity.

Other archival systems use offline data storage in which archive data are written to tape or other removable media using data archiving software rather than being kept online. Because tape can be ____7____, tape-based archives consume far less power than disk systems. This translates to lower archive storage costs.

Cloud storage is another possible archive target. Amazon Glacier, for example, is designed for data archiving. Cloud storage is inexpensive but requires an ongoing ___8___. In addition, costs can grow over time as more data are added to the cloud archive.

The archival process is almost always automated using archiving software. The capabilities of such software vary from one vendor to the next, but generally speaking, the software will ___9___ move aging data to the archives according to a data archival policy set by the storage administrator. This policy may also include specific retention requirements for each type of data. Some archiving software will automatically purge data from the archives once they have exceeded the lifespan mandated by the organization's data retention policy. Many backup software platforms are adding archiving functionality to their ___10___. Depending on your needs, this can be a cost-effective and efficient way to archive data. However, these products may not include all of the functionality found in a dedicated archive software product.

Text B
Cloud Storage

扫码听课文

Cloud storage is a model of computer data storage in which the digital data are stored in logical pools. The physical storage spans multiple servers (sometimes in multiple locations), and the physical environment is typically owned and managed by a hosting company. These cloud storage providers are responsible for keeping the data available and accessible, and the physical environment protected and running. People and organizations buy or lease storage capacity from the providers to store user, organization, or application data.

Cloud storage services may be accessed through a colocated cloud computing service, a web service application programming interface (API) or by applications that utilize the API, such as cloud desktop storage, a cloud storage gateway or Web-based content management systems.

Cloud storage is based on highly virtualized infrastructure and is like broader cloud computing in terms of accessible interfaces, near-instant elasticity and scalability, multi-tenancy, and metered resources. Cloud storage services can be utilized from an off-premises service (Amazon S3) or deployed on-premises (ViON Capacity Services).

Cloud storage typically refers to a hosted object storage service, but the term has broadened to include other types of data storage that are now available as a service, like block storage.

Object storage services like Amazon S3, Oracle Cloud Storage and Microsoft Azure Storage, object storage software like Openstack Swift, object storage systems like EMC Atmos, EMC ECS and Hitachi Content Platform, and distributed storage research projects like OceanStore and VISION Cloud are all examples of storage that can be hosted and deployed with cloud storage characteristics.

Cloud storage is:

- Made up of many distributed resources, but still acts as one, either in a federated or a cooperative storage cloud architecture.
- Highly fault tolerant through redundancy and distribution of data.
- Highly durable through the creation of versioned copies.
- Typically eventually consistent with regard to data replicas.

1. Advantages

Companies need only pay for the storage they actually use, typically an average of consumption during a month. This does not mean that cloud storage is less expensive, only that it incurs operating expenses rather than capital expenses.

Businesses using cloud storage can cut their energy consumption by up to 70%, making them a more green business. Also at the vendor level they are dealing with higher levels of energy so they will be more equipped with managing it in order to keep their own costs down as well.

Organizations can choose between off-premises and on-premises cloud storage options, or a mixture of the two options, depending on relevant decision criteria that are complementary to initial direct cost savings potential; for instance, continuity of operations (COOP), disaster recovery (DR), security, and records retention laws, regulations, and policies.

Storage availability and data protection are intrinsic to object storage architecture, so depending on the application, the additional technology, effort and cost to add availability and protection can be eliminated.

Storage maintenance tasks, such as purchasing additional storage capacity, are offloaded to the responsibility of a service provider.

Cloud storage provides users with immediate access to a broad range of resources and applications hosted in the infrastructure of another organization via a web service interface.

Cloud storage can be used for copying virtual machine images from the cloud to on-premises locations or to import a virtual machine image from an on-premises location to the cloud image library. In addition, cloud storage can be used to move virtual machine images between user accounts or between data centers.

Cloud storage can be used as natural disaster proof backup, normally there are 2 or 3 different backup servers located in different places around the globe.

Cloud storage can be mapped as a local drive with the WebDAV protocol. It can function as a central file server for organizations with multiple office locations.

2. Potential Concerns

2.1 Attack Surface Area

Outsourcing data storage increases the attack surface area.

When data have been distributed they are stored at more locations, increasing the risk of unauthorized physical access to the data. For example, in cloud based architecture, data are replicated and moved frequently so the risk of unauthorized data recovery increases dramatically. Such as in the case of disposal of old equipment, reuse of drives, reallocation of storage space. The manner that data are replicated depends on the service level a customer chooses and on the service provided. When encryption is in place it can ensure confidentiality. Crypto-shredding can be used when disposing of data (on a disk).

The number of people with access to the data who could be damaged (e.g., bribed, or coerced) increases dramatically. A single company might have a small team of administrators, network engineers, and technicians. But a cloud storage company will have many customers and thousands of servers. Therefore a much larger team of technical staff with physical and electronic access to almost all of the data at the entire facility or perhaps the entire company. Decryption keys that are kept by the service user, as opposed to the service provider, limit the access to data by service provider employees. As for sharing multiple data in the cloud with multiple users, a large number of keys have to be distributed to users via secure channels for decryption, also they have to be securely stored and managed by the users in their devices. Storing these keys requires rather expensive secure storage. To overcome that, key-aggregate cryptosystem can be used.

It increases the number of networks over which the data travel. Instead of just a local area network (LAN) or storage area network (SAN), data stored on a cloud require a WAN (wide area network) to connect them both.

By sharing storage and networks with many other users/customers it is possible for other customers to access your data. Sometimes because of erroneous actions, faulty equipment, a bug and sometimes because of criminal intent. This risk applies to all types of storage and not only cloud storage. The risk of having data read during transmission can be mitigated through encryption technology. Encryption in transit protects data as they are being transmitted to and from the cloud service. Encryption at rest protects data that are stored at the service provider. Encrypting data in an on-premises cloud service on-ramp system can provide both kinds of encryption protection.

2.2 Supplier Stability

Companies are not permanent and the services and products they provide can change. Outsourcing data storage to another company needs careful investigation and nothing is ever certain. Contracts set in stone can be worthless when a company ceases to exist or its circumstances change. Companies can:

- Go bankrupt.
- Expand and change their focus.
- Be purchased by other larger companies.
- Be purchased by a company headquartered in or move to a country that negates compliance

with export restrictions and thus necessitates a move.

- Suffer an irrecoverable disaster.

2.3 Accessibility

Performance for outsourced storage is likely to be lower than local storage, depending on how much a customer is willing to spend for WAN bandwidth.

Reliability and availability depend on wide area network availability and on the level of precautions taken by the service provider. Reliability should be based on hardware as well as various algorithms used.

2.4 Other Concerns

Security of stored data and data in transit may be a concern when storing sensitive data at a cloud storage provider

Users with specific records-keeping requirements, such as public agencies that must retain electronic records according to statute, may encounter complications with using cloud computing and storage. For instance, the U.S. Department of Defense designated the Defense Information Systems Agency (DISA) to maintain a list of records management products that meet all of the records retention, personally identifiable information (PII), and security (Information Assurance, IA) requirements.

Cloud storage is a rich resource for both hackers and national security agencies. Because the cloud holds data from many different users and organizations, hackers see them as a very valuable target.

The legal aspect, from a regulatory compliance standpoint, is of concern when storing files domestically and especially internationally.

New Words

pool	[puːl]	*n.* 池
server	[ˈsɜːvə]	*n.* 服务器
location	[ləʊˈkeɪʃn]	*n.* 位置，场所
environment	[ɪnˈvaɪrənmənt]	*n.* 环境，外界；工作平台
responsible	[rɪˈspɒnsəbl]	*adj.* 尽责的，负有责任的
available	[əˈveɪləbl]	*adj.* 可利用的；可获得的；能找到的
accessible	[əkˈsesəbl]	*adj.* 可访问的
utilize	[ˈjuːtɪlaɪz]	*vt.* 利用，使用
gateway	[ˈgeɪtweɪ]	*n.* 门；入口；途径
virtualized	[ˈvɜːtʃʊəlaɪzd]	*adj.* 虚拟化的
infrastructure	[ˈɪnfrəstrʌktʃə]	*n.* 基础设施；基础建设

elasticity	[ˌiːlæ'stɪsɪtɪ]	*n.* 弹性；灵活性；伸缩性
scalability	[skeɪlə'bɪlɪtɪ]	*n.* 可扩展性；可伸缩性；可量测性
tenancy	['tenənsɪ]	*n.* 租用，租赁；租期
distributed	[dɪs'trɪbjuːtɪd]	*adj.* 分布式的
federate	['fedəreɪt]	*v.*（使）结成联盟
cooperative	[kəʊ'ɒpərətɪv]	*adj.* 合作的；协助的；共同的
durable	['djʊərəbl]	*adj.* 耐用的，耐久的；持久的；长期的
		n. 耐用品，耐久品
replica	['replɪkə]	*n.* 复制品
consumption	[kən'sʌmpʃn]	*n.* 消费
expensive	[ɪk'spensɪv]	*adj.* 昂贵的，花钱多的
expense	[ɪk'spens]	*n.* 费用；消耗
		vt. 向……收取费用
off-premise	['ɔːfpr'emɪs]	*adj.* 外部部署的
on-premise	[ɒn'premɪs]	*adj.* 本地部署的
regulation	[ˌregju'leɪʃn]	*n.* 规章；规则
		adj. 规定的
intrinsic	[ɪn'trɪnsɪk]	*adj.* 固有的，内在的，本质的
maintenance	['meɪntənəns]	*n.* 维持，保持；保养；维护；维修
immediate	[ɪ'miːdɪət]	*adj.* 立即的；直接的
outsourcing	['aʊtsɔːsɪŋ]	*n.* 外包，外购
unauthorized	[ʌn'ɔːθəraɪzd]	*adj.* 未经授权的；未经许可的；未经批准的
dramatically	[drə'mætɪklɪ]	*adv.* 显著地，剧烈地
disposal	[dɪ'spəʊzl]	*n.*（事情的）处置；清理
		adj. 处理（或置放）废品的
reuse	[ˌriː'juːz]	*vt.* 重用，复用
reallocation	[ˌriːˌælə'keɪʃn]	*vt.* 再分配
encryption	[ɪn'krɪpʃn]	*n.* 编码，加密
confidentiality	[ˌkɒnfɪˌdenʃi'ælɪtɪ]	*n.* 机密性
crypto-shredding	['krɪptəʊ-'ʃredɪŋ]	*n.* 密码粉碎
dispose	[dɪ'spəʊz]	*v.* 处理，处置；安排
damage	['dæmɪdʒ]	*vt.* 损害
bribe	[braɪb]	*v.* 贿赂，行贿
		n. 贿赂
coerce	[kəʊ'ɜːs]	*vt.* 控制，限制；威胁；逼迫
decryption	[diː'krɪpʃn]	*n.* 解密，译码
channel	['tʃænl]	*n.* 通道，渠道

erroneous	[ɪ'rəʊnɪəs]	*adj.* 错误的，不正确的
bug	[bʌg]	*n.* 缺陷，漏洞
stability	[stə'bɪlɪtɪ]	*n.* 稳定（性）；稳固
investigation	[ɪnˌvestɪ'geɪʃn]	*n.* 调查，研究
bankrupt	['bæŋkrʌpt]	*adj.* 破产的，倒闭的
		n. 破产者
		vt. 使破产
headquarter	['hed'kwɔːtə]	*vi.* 设总部
		vt. 将……的总部设在；把……放在总部里
irrecoverable	[ˌɪrɪ'kʌvərəbl]	*adj.* 无可挽救的；不可弥补
precaution	[prɪ'kɔːʃn]	*n.* 预防，防备，警惕；预防措施
		vt. 使提防
hardware	['hɑːdweə]	*n.* 硬件
multiplicity	[ˌmʌltɪ'plɪsɪtɪ]	*n.* 多样性
transit	['trænzɪt]	*vt.* 传输
agency	['eɪdʒənsɪ]	*n.* 代理；机构
encounter	[ɪn'kaʊntə]	*vt.* 不期而遇
complication	[ˌkɒmplɪ'keɪʃn]	*n.* 纠纷；混乱
hacker	['hækə]	*n.* 黑客
standpoint	['stændpɔɪnt]	*n.* 立场，观点
domestically	[də'mestɪklɪ]	*adv.* 国内地；适合国内地
internationally	[ˌɪntə'næʃnəlɪ]	*adv.* 国际性地，国际上地，国际间地

Phrases

cloud storage	云存储
physical storage	物理存储
colocated cloud computing	协同云计算
cloud desktop	云桌面
content management system	内容管理系统
be based on...	基于……
block storage	块存储
object storage	对象存储
distributed storage	分布式存储
cooperative storage cloud	协同存储云
fault tolerant	容错

eventually consistent	最终一致性
data protection	数据保护
virtual machine image	虚拟机映像
natural disaster proof backup	防自然灾难备份
local drive	本地驱动器
storage space	存储空间
key-aggregate cryptosystem	密钥聚合密码系统
faulty equipment	设备故障
on-ramp system	入站匝道系统
set in stone	一成不变
export restriction	出口限制
sensitive data	敏感数据

Abbreviations

API（Application Programming Interface）	应用程序接口
COOP（Continuity of Operation）	连续性经营
DR（Disaster Recovery）	灾难恢复
LAN（Local Area Network）	局域网
SAN（Storage Area Network）	存储域网
WAN（Wide Area Network）	广域网
DISA（Defense Information Systems Agency）	国防信息系统局
PII（Personally Identifiable Information）	个人身份信息
IA（Information Assurance）	信息保障

Exercises

【Ex.5】Answer the following questions according to the text.

1. How may cloud storage services be accessed?
2. What does cloud storage typically refer to?
3. What is cloud storage made up of?
4. How can organizations store their data on cloud?
5. What can cloud storage be used for?
6. When data have been distributed, where are they stored?
7. What has to be done as for sharing multiple data in the cloud with multiple users?
8. What does outsourcing data storage to another company need? When can contracts set in stone

be worthless?

9. What do reliability and availability depend on?

10. For whom is cloud storage a rich resource? Why do hackers see it as a very valuable target?

参考译文

数据仓库

在计算技术中，数据仓库（DW），也称为企业数据仓库（EDW），是用于报告和数据分析的系统，并且被认为是商业智能的核心组件。数据仓库是来自一个或多个不同来源的集成数据的中心存储库。它们将当前数据和历史数据存储在一起，用于为整个企业中的工作人员创建分析报告。

存储在仓库中的数据从运营系统上载（如营销或销售数据）。数据在被存储之后、在用于数据仓库报告之前，可能需要进行清理等附加操作以确保数据质量（见图 7-1）。

典型的基于提取、转换、加载（ETL）的数据仓库使用临时存储、数据集成和访问层来放置其关键功能。临时层或临时数据库存储从各个不同的源数据系统提取的原始数据。集成层通过转换来自临时层的数据来集成不同的数据集，将这些转换了的数据存储在操作数据存储（ODS）数据库中，然后将集成数据移动到另一个数据库，通常称为数据仓库数据库。访问层可帮助用户检索数据。

数据经过清理、转换、编目，可供管理人员和其他业务专业人员用于数据挖掘、在线分析处理、市场研究和决策支持。但是，检索和分析数据，提取、转换和加载数据以及管理数据字典的方法也被认为是数据仓库系统的重要组成部分。数据仓库的这种宽泛使用更常见。因此，数据仓库的扩展定义包括商业智能工具，提取、转换和加载数据到存储库的工具，以及管理和检索元数据的工具。

1. 优点

数据仓库保存源事务系统的信息副本。这种架构复杂性提供了以下机会。

- 将来自多个源的数据集成到单个数据库和数据模型中。给单个数据库提供更多的数据，这样就可以使用单个查询引擎在操作性数据中显示数据。
- 减轻了由于尝试在事务处理数据库中运行大型、长时间的分析查询而导致的事务处理系统中的数据库隔离级别锁争用问题。
- 保存数据历史记录，即使源事务系统没有保存。
- 集成来自多个源系统的数据，实现整个企业的中央视图。这种好处总是有价值的，尤其在组织通过合并而增长时有用。
- 通过提供一致的代码和描述、标记甚至修复错误数据来提高数据质量。
- 始终展示组织的信息。

- 为所有感兴趣的相关数据提供单一的通用数据模型，而不管数据的来源如何。
- 重构数据，使其对业务用户有意义。
- 重构数据，使其即使对于复杂的分析查询也能提供出色的查询性能，并且不会影响操作系统。
- 为运营业务应用程序增加价值，特别是客户关系管理（CRM）系统。
- 使决策支持查询更容易编写。
- 组织数据并消除重复数据的歧义。

2. 通用环境

数据仓库和市场的环境包括以下内容。

- 向仓库或集市提供数据的源系统。
- 数据集成技术以及准备所需数据的流程。
- 用于在组织的数据仓库或数据集市中存储数据的不同体系结构。
- 适用于各种用户的不同工具和应用程序。
- 必须建立元数据、数据质量和治理流程，以确保仓库或市场满足其目的。

关于上面列出的源系统，R. Kelly Rainer 表示，“数据仓库中数据的一个共同来源是公司的运营数据库，它可以是关系数据库”。

关于数据集成，Rainer 表示，“有必要从源系统中提取数据，转换它们，并将它们加载到数据集市或仓库中”。

如今，最成功的公司是能够快速灵活地应对市场变化和机遇的公司。这种响应的关键是分析师和管理人员有效和高效地使用数据和信息。“数据仓库”是历史数据的存储库，这些按主题组织好，以便为决策者提供支持。一旦数据存储在数据集市或仓库中，就可以访问它们。

3. 相关系统（数据集市、OLAP、OLTP、预测分析）

数据集市是数据仓库的一种简单形式，专注于单个主题（或功能区域），因此它们从有限数量的来源（如销售、财务或营销）中提取数据。数据集市通常由组织内的单个部门构建和控制。来源可以是内部操作系统、中央数据仓库或外部数据。反规范化是该系统中数据建模技术的标准。鉴于数据集市通常仅是数据仓库的一个数据子集，这通常更容易和更快实现。

数据仓库和数据集市的区别

属性	数据仓库	数据集市
数据范围	企业范围	部门范围
主题范围数目	多个	单个
建立难易	难	易
建立所需时间	长	短
所需内存	大	有限

数据集市的类型包括依赖、独立和混合数据集市。

联机分析处理（OLAP）的特点是业务量相对较低。查询通常非常复杂并涉及聚合。对于 OLAP 系统，响应时间是一种有效性度量。数据挖掘技术广泛使用 OLAP 应用程序。OLAP 数据库在多维模式（通常是星型模式）中存储聚合的历史数据。OLAP 系统通常具有几个小时的数据延迟，而数据集市中延迟预计接近一天。OLAP 方法用于分析来自多个来源和视角的多维数据。OLAP 中的三个基本操作是：上卷（合并）、下钻以及切片和切块。

联机事务处理（OLTP）的特点是大量短时的在线事务（插入、更新、删除）。OLTP 系统强调在多访问环境中非常快速的查询处理和维护数据完整性。对于 OLTP 系统，有效性通过每秒的事务数来衡量。OLTP 数据库包含详细的当前数据。用于存储事务数据库的模式是实体模型（通常为 3NF）。规范化是该系统中数据建模技术的标准。

预测分析是使用可用于预测未来结果的复杂数学模型来查找和量化数据中的隐藏模式。预测分析与 OLAP 的不同之处在于，OLAP 侧重于历史数据分析，并且本质上是反应性的，而预测分析则侧重于未来。这些系统还用于客户关系管理（CRM）。

4. 设计方法

4.1 自下而上的设计

在自下而上的设计中，首先创建数据集市以提供特定业务流程的报告和分析功能，然后可以集成这些数据集市来创建综合数据仓库。

4.2 自上而下的设计

自上而下的方法使用标准化的企业数据模型设计。“原子”数据，即最详细程度的数据，存储在数据仓库中。包含特定业务流程或特定部门所需数据的维度数据集市是从数据仓库创建出来的。

4.3 混合设计

数据仓库（DW）通常类似于中心辐射型架构。给仓库提供数据的老系统通常包括客户关系管理和企业资源规划，它们会生成大量数据。为了整合这些不同的数据模型，并促进提取转换加载过程，数据仓库通常使用操作数据存储，其中的信息被解析为实际的 DW。为了减少数据冗余，较大的系统通常以标准化方式存储数据。之后，可以在数据仓库之上构建特定报告的数据集市。

5. 数据仓库的特征

有一些基本功能可以定义数据仓库中的数据，包括主题方向、数据集成、时变、非易失性数据和数据粒度。

5.1 面向主题

与操作系统不同，数据仓库中的数据围绕企业主题（数据库规范化）。主题方向对决策非常有用，收集所需对象称为面向主题。

5.2 集成

数据仓库中的数据已集成。由于它来自多个操作系统，因此必须删除所有不一致的内容。一致性包括命名约定、变量度量、编码结构、数据的物理属性等。

5.3 时变

虽然因为操作系统支持日常操作而反映当前值，但数据仓库中的数据代表长时间（长达10年）的数据，这意味着它存储历史数据。它主要用于数据挖掘和预测。如果用户正在搜索特定客户的购买模式，则用户需要查看当前和过去购买的数据。

5.4 非易失性

数据仓库中的数据是只读的，这意味着无法更新、创建或删除它。

5.5 总结

在数据仓库中，数据在不同级别汇总。用户可以先查看整个区域中产品的总销售单位，然后再查看该区域中的状态。最后，他们可以检查某个状态下的个体商店。因此，通常情况下，分析从较高级别开始，然后向下移动到较低级别的详细信息。

Unit 8

Text A
Data Processing (1)

1. Data Pre-Processing

Data pre-processing is an important step in the data mining process. The phrase "garbage in, garbage out" is particularly applicable to data mining and machine learning projects. Data gathering methods are often loosely controlled, resulting in out-of-range values (e.g., Income: −100), impossible data combinations (e.g., Sex: Male, Pregnant: Yes), missing values, etc. Analyzing data that has not been carefully screened for such problems can produce misleading results. Thus, the representation and quality of data is first and foremost before running an analysis. Often, data pre-processing is the most important phase of a machine learning project, especially in computational biology.

If there is much irrelevant and redundant information present or noisy and unreliable data, then knowledge discovery during the training phase is more difficult. Data preparation and filtering steps can take considerable amount of processing time. Data pre-processing includes cleaning, instance selection, normalization, transformation, feature extraction and selection, etc. The product of data pre-processing is the final training set.

2. Data Editing

Data editing is defined as the process involving the review and adjustment of collected survey data. The purpose is to control the quality of the collected data. Data editing can be performed manually, with the assistance of a computer or a combination of both. The methods of data editing mainly include the following types.

2.1 Interactive Editing

The term interactive editing is commonly used for modern computer-assisted manual editing. Most interactive data editing tools applied at National Statistical Institutes (NSI) allow one to check the specified edits during or after data entry, and if necessary to correct erroneous data immediately. Several approaches can be followed to correct erroneous data:

- Recontact the respondent.
- Compare the respondent's data to his data from previous year.
- Compare the respondent's data to data from similar respondents.
- Use the subject matter knowledge of the human editor.

Interactive editing is a standard way to edit data. It can be used to edit both categorical and continuous data. Interactive editing reduces the time frame needed to complete the cyclical process of review and adjustment.

2.2 Selective Editing

Selective editing is an umbrella term for several methods to identify the influential errors, and outliers. Selective editing techniques aim to apply interactive editing to a well-chosen subset of the records, such that the limited time and resources available for interactive editing are allocated to those records where they have the most effect on the quality of the final estimates of publication figures. In selective editing, data are split into two streams:

- The critical stream.
- The non-critical stream.

The critical stream consists of records that are more likely to contain influential errors. These critical records are edited in a traditional interactive manner. The records in the non-critical stream which are unlikely to contain influential errors are not edited in a computer assisted manner.

2.3 Macro Editing

There are two methods of macro editing:

- Aggregation method

This method is followed in almost every statistical agency before publication: Verifying whether figures to be published seem plausible. This is accomplished by comparing quantities in publication tables with same quantities in previous publications. If an unusual value is observed, a micro editing procedure is applied to the individual records and fields contributing to the suspicious quantity.

- Distribution method

Data available are used to characterize the distribution of the variables. Then all individual values are compared with the distribution. Records containing values that could be considered uncommon (given the distribution) are candidates for further inspection and possibly for editing.

2.4 Automatic Editing

In automatic editing records are edited by a computer without human intervention. Prior knowledge on the values of a single variable or a combination of variables can be formulated as a set of edit rules which specify or constrain the admissible values.

3. Data Reduction

Data reduction is the transformation of numerical or alphabetical digital information derived empirically or experimentally into a corrected, ordered, and simplified form. The basic concept is the reduction of multitudinous amounts of data down to the meaningful parts.

When information is derived from instrument readings there may also be a transformation from analog to digital form. When the data are already in digital form the "reduction" of the data typically involves some editing, scaling, encoding, sorting, collating, and producing tabular summaries. When the observations are discrete but the underlying phenomenon is continuous then smoothing and interpolation are often needed. Often the data reduction is undertaken in the presence of reading or measurement errors. Some idea of the nature of these errors is needed before the most likely value may be determined.

These are common techniques used in data reduction:

- Order by some aspect of size.
- Table diagonalization, whereby rows and columns of tables are re-arranged to make patterns easier to see.
- Round drastically to one, or at most two, effective digits.
- Use averages to provide a visual focus as well as a summary.
- Use layout and labeling to guide the eye.
- Remove chart junk, such as pictures and lines.
- Give a brief verbal summary.

4. Data Wrangling

Data wrangling, sometimes referred to as data munging, is the process of transforming and mapping data from one "raw" data form into another format with the intent of making them more appropriate and valuable for a variety of downstream purposes such as analytics. A data wrangler is a person who performs these transformation operations.

This may include further munging, data visualization, data aggregation, training a statistical model, as well as many other potential uses. Data munging as a process typically follows a set of general steps which begin with extracting the data in a raw form from the data source, "munging" the raw data using algorithms (e.g., sorting) or parsing the data into predefined data structures, and finally depositing the resulting content into a data sink for storage and future use.

4.1 Typical Use

The data transformations are typically applied to distinct entities (e.g. fields, rows, columns, data values, etc.) within a data set, and could include such actions as extractions, parsing, joining, standardizing, augmenting, cleansing, consolidating and filtering to create desired wrangling outputs that can be leveraged downstream.

The recipients could be individuals, such as data architects or data scientists who will investigate the data further, business users who will consume the data directly in reports, or systems that will further process the data and write them into targets such as data warehouses, data lakes or downstream applications.

4.2 Modus Operandi

Depending on the amount and format of the incoming data, data wrangling has traditionally been performed manually (e.g., via spreadsheets such as Excel) or via hand-written scripts in languages such as Python or SQL. R, a language often used in data mining and statistical data analysis, is now also often used for data wrangling. Other terms for these processes have included data franchising, data preparation and data munging.

New Words

applicable	[ə'plɪkəbl]	*adj.* 适当的；可应用的
gather	['gæðə]	*vt.* 收集，采集
		vi. 逐渐增加，积聚
		n. 聚集
out-of-range	['aʊt əv 'reɪndʒ]	*adj.* 溢出的；出界的
impossible	[ɪm'pɒsəbl]	*adj.* 不可能的，做不到的
screen	[skriːn]	*vt.* 筛选
		n. 屏幕
misleading	[ˌmɪs'liːdɪŋ]	*adj.* 误导性的；引入歧途的
representation	[ˌreprɪzen'teɪʃn]	*n.* 表现；陈述；表现……的事物
foremost	['fɔːməʊst]	*adj.* 最初的，最前面的
		adv. 首先，第一
phase	[feɪz]	*n.* 阶段
computational	[ˌkɒmpjʊ'teɪʃənl]	*adj.* 计算的
irrelevant	[ɪ'reləvənt]	*adj.* 不相干的
noisy	['nɔɪzɪ]	*adj.* 充满噪声的
		n. 噪声
unreliable	[ˌʌnrɪ'laɪəbl]	*adj.* 不可靠的
preparation	[ˌprepə'reɪʃn]	*n.* 准备，预备；准备工作
filter	['fɪltə]	*v.* 过滤
considerable	[kən'sɪdərəbl]	*adj.* 相当大（或多）的；该注意的，应考虑的
selection	[sɪ'lekʃn]	*n.* 选择

transformation	[ˌtrænsfəˈmeɪʃn]	*n.* 变化；转换
extraction	[ɪkˈstrækʃn]	*n.* 抽出；提取
perform	[pəˈfɔːm]	*v.* 执行，履行
manually	[ˈmænjʊəlɪ]	*adv.* 用手地，手动地
assistance	[əˈsɪstəns]	*n.* 帮助，援助
interactive	[ˌɪntərˈæktɪv]	*adj.* 互相作用的，相互影响的；互动的
recontact	[ˌrɪˈkɒntækt]	*v.* 重新联系
respondent	[rɪˈspɒndənt]	*n.* 应答者
categorical	[ˌkætəˈgɒrɪkl]	*adj.* 分类的，按类别的
continuous	[kənˈtɪnjʊəs]	*adj.* 连续的；延伸的；不断的
frame	[freɪm]	*n.* 框架
cyclical	[ˈsɪklɪkəl]	*adj.* 循环的；周期的；环状的
identify	[aɪˈdentɪfaɪ]	*vt.* 识别，认出；确定 *vi.* 确定；认同
influential	[ˌɪnfluˈenʃl]	*adj.* 有影响的
outlier	[ˈaʊtlaɪə]	*n.* 离群值；异常值
estimate	[ˈestɪmɪt] [ˈestɪmeɪt]	*n.* 估计，预测 *vt.* 估计，估算；评价，评论
stream	[striːm]	*n.*（数据）流 *vt.* & *vi.* 流，流动
critical	[ˈkrɪtɪkl]	*adj.* 关键的；极重要的
macro	[ˈmækrəʊ]	*n.* 宏 *adj.* 巨大的
unusual	[ʌnˈjuːʒʊəl]	*adj.* 不常见的，不普通的，难得的，罕有的；异乎寻常的
uncommon	[ʌnˈkɒmən]	*adj.* 不寻常的；罕见的
inspection	[ɪnˈspekʃn]	*n.* 检查；检验
constrain	[kənˈstreɪn]	*vt.* 限制；约束
admissible	[ədˈmɪsəbl]	*adj.* 可容许的
reduction	[rɪˈdʌkʃn]	*n.* 精简；减少；降低
empirically	[ɪmˈpɪrɪklɪ]	*adv.* 以经验为主地
experimentally	[ɪkˌsperɪˈmentəlɪ]	*adv.* 实验（性）地；实际上，通过实验
concept	[ˈkɒnsept]	*n.* 观念，概念
multitudinous	[ˌmʌltɪˈtjuːdɪnəs]	*adj.* 大量的，多种多样的
meaningful	[ˈmiːnɪŋfl]	*adj.* 有意义的
instrument	[ˈɪnstrəmənt]	*n.* 仪器；手段，工具
analog	[ˈænəlɔːg]	*n.* 模拟 *adj.* 模拟的

sort	[sɔːt]	*n.* 分类 *v.* 排序，分类
collate	[kəˈleɪt]	*vt.* 核对，校对；检查
tabular	[ˈtæbjʊlə]	*adj.* 表格的；按表格计算的
observation	[ˌɒbzəˈveɪʃn]	*n.* 观察；观察力
discrete	[dɪˈskriːt]	*adj.* 分离的，不相关联的
phenomenon	[fəˈnɒmɪnən]	*n.* 现象，事件
smooth	[ˈsmuːð]	*v.*（使）光滑，（使）平坦
interpolation	[ɪnˌtɜːpəˈleɪʃn]	*n.* 插补
undertake	[ˌʌndəˈteɪk]	*vt.* 保证；同意，答应；承诺
diagonalization	[daɪˌægənəlaɪˈzeɪʃn]	*n.* 对角化，对角线化
drastically	[ˈdrɑːstɪklɪ]	*adv.* 大大地，彻底地
layout	[ˈleɪaʊt]	*n.* 布局，安排，设计；布置图，规划图
wrangle	[ˈræŋgl]	*v.* 整理
downstream	[ˌdaʊnˈstriːm]	*adv.* 在下游地；顺流地 *adj.* 顺流而下的；在下游方向的
wrangler	[ˈræŋglə]	*n.* 整理者
potential	[pəˈtenʃl]	*adj.* 潜在的，有可能的 *n.* 潜力，潜能
parse	[pɑːz]	*vt.* 分析，解析
predefined	[priːdɪˈfaɪnd]	*adj.* 预定义的
deposit	[dɪˈpɒzɪt]	*n.* 沉淀物 *v.* 沉淀
join	[dʒɔɪn]	*v.* 连接；联结
investigate	[ɪnˈvestɪgeɪt]	*vt.* 调查；审查；研究 *vi.* 作调查

Phrases

data pre-processing	数据预处理
garbage in, garbage out	垃圾进，垃圾出
data gathering	数据收集，数据采集
missing value	缺失值
computational biology	计算生物学
knowledge discovery	知识发现
training set	训练集

survey data	调查数据
be split into	被分为
macro editing	宏编辑
aggregation method	聚合方法
distribution method	分配方法
be compared with ...	与……比较
be formulated as	被表示为
data reduction	数据精简
be derived from	由……而来，起源于
analog form	模拟形式
digital form	数字形式
chart junk	图表垃圾
data wrangling	数据整理
data munging	数据调整
data visualization	数据可视化
data aggregation	数据聚合
data franchising	数据特许权
data preparation	数据准备

Abbreviations

NSI（National Statistical Institutes）　　国家统计局

Exercises

【Ex.1】Answer the following questions according to the text.

1. What does data pre-processing include? What is its product?
2. What is data editing defined as? What is its purpose? How can it be performed?
3. What approaches can be followed to correct erroneous data?
4. What is selective editing? What does it aim to do?
5. How many methods are there of macro editing? What are they?
6. What is data reduction? What is the basic concept?
7. What does the "reduction" of the data typically involve when the data has been already in digital form?
8. What is data wrangling?
9. What are the data transformations typically applied to?
10. What is R often used in?

【Ex.2】Translate the following terms or phrases from English into Chinese and vice versa.

1. aggregation method	1.
2. extraction	2.
3. data preparation	3.
4. data reduction	4.
5. data visualization	5.
6. 数据整理	6.
7. 数据收集，数据采集	7.
8. 知识发现	8.
9. *adj.*循环的；周期的；环状的	9.
10. *v.*过滤	10.

【Ex.3】Translate the following passages into Chinese.

KDD

Knowledge discovery in databases (KDD) is the process of discovering useful knowledge from a collection of data. This widely used data mining technique is a process that includes data preparation and selection, data cleansing, incorporating prior knowledge on data sets and interpreting accurate solutions from the observed results.

Major KDD application areas include marketing, fraud detection, telecommunication and manufacturing.

Traditionally, data mining and knowledge discovery were performed manually. As time passed, the amount of data in many systems grew to larger than terabyte size, and could no longer be maintained manually. Moreover, for the successful existence of any business, discovering underlying patterns in data is considered essential. As a result, several software tools were developed to discover hidden data and make assumptions, which formed a part of artificial intelligence.

The KDD process has reached its peak in the last 10 years. It now houses many different approaches to discovery, which include inductive learning, Bayesian statistics, semantic query optimization, knowledge acquisition for expert systems and information theory. The ultimate goal is to extract high-level knowledge from low-level data.

【Ex.4】Fill in the blanks with the words given below.

overload	ranking	matched	specific	objects
engines	identify	retrieval	represented	statements

Information Retrieval

Information retrieval (IR) is the activity of obtaining information system resources relevant to an information need from a collection of information resources. Searches can be based on full-text or other content-based indexing. Information ___1___ is the science of searching for information in a document, searching for documents themselves, and also searching for metadata that describe data, and for databases of texts, images or sounds.

Automated information retrieval systems are used to reduce what has been called information ___2___. An IR system is a software that provides access to books, journals and other documents, stores them and manages the document. Web search ___3___ are the most visible IR applications.

An information retrieval process begins when a user enters a query into the system. Queries are formal ___4___ of information needs, for example search strings in web search engines. In information retrieval a query does not uniquely ___5___ a single object in the collection. Instead, several objects may match the query, perhaps with different degrees of relevancy.

An object is an entity that is represented by information in a content collection or database. User queries are ___6___ against the database information. However, as opposed to classical SQL queries of a database, in information retrieval the results returned may or may not match the query, so results are typically ranked. This ___7___ of results is a key difference of information retrieval searching compared to database searching.

Depending on the application, the data objects may be, for example, text documents, images, audio, mind maps or videos. Often the documents themselves are not kept or stored directly in the IR system, but are instead ___8___ in the system by document surrogates or metadata.

Most IR systems compute a numeric score on how well each object in the database matches the query, and rank the objects according to this value. The top ranking ___9___ are then shown to the user. The process may then be iterated if the user wishes to refine the query.

For effectively retrieving relevant documents by IR strategies, the documents are typically transformed into a suitable representation. Each retrieval strategy incorporates a ___10___ model for its document representation purposes. The picture on the right illustrates the relationship of some common models. In the picture, the models are categorized according to two dimensions: The mathematical basis and the properties of the model.

Text B
Data Processing (2)

扫码听课文

1. Data Scraping

Data scraping is a technique in which a computer program extracts data from human-readable output coming from another program.

1.1 Description

Normally, data transfer between programs is accomplished using data structures suited for automated processing by computers, not people. Such interchange formats and protocols are typically rigidly structured, well-documented, easily parsed, and keep ambiguity to a minimum. Very often, these transmissions are not human-readable at all.

Thus, the key element that distinguishes data scraping from regular parsing is that the output being scraped is intended for display to an end user, rather than as input to another program, and is therefore usually neither documented nor structured for convenient parsing. Data scraping often involves ignoring binary data (usually images or multimedia data), display formatting, redundant labels, superfluous commentary, and other information which is either irrelevant or hinders automated processing.

Data scraping is most often done either to interface to a legacy system which has no other mechanism which is compatible with current hardware, or to interface to a third-party system which does not provide a more convenient API. In the second case, the operator of the third-party system will often see screen scraping as unwanted, due to reasons such as increased system load, the loss of advertisement revenue, or the loss of control of the information content.

Data scraping is generally considered an ad hoc inelegant technique, often used only as a "last resort" when no other mechanism for data interchange is available. Aside from the higher programming and processing overhead, output displays intended for human consumption often change structure frequently. Humans can cope with this easily, but a computer program may report nonsense, having been told to read data in a particular format or from a particular place, and with no knowledge of how to check its results for validity.

1.2 Technical Variants

1.2.1 Screen Scraping

Screen scraping is normally associated with the programmatic collection of visual data from a source, instead of parsing data as in Web scraping. Originally, screen scraping referred to the practice of reading text data from a computer display terminal's screen. This was generally done by reading the terminal's memory through its auxiliary port, or by connecting the terminal output port of one computer system to an input port on another.

The term screen scraping is also commonly used to refer to the bidirectional exchange of data. This could be the simple cases where the controlling program navigates through the user interface, or more complex scenarios where the controlling program is entering data into an interface meant to be used by a human.

1.2.2 Web Scraping

Web pages are built using text-based mark-up languages (HTML and XHTML), and frequently contain a wealth of useful data in text form. However, most web pages are designed for human end users and not for ease of automated use. Because of this, tool kits that scrape web content are created. A web scraper is an API or tool to extract data from a web site. Companies like Amazon AWS and Google provide web scraping tools, services and public data available free of cost to end users. Newer forms of web scraping involve listening to data feeds from web servers. For example, JSON is commonly used as a transport storage mechanism between the client and the web server.

Recently, companies have developed web scraping systems that rely on using techniques in DOM parsing, computer vision and natural language processing to simulate the human processing that occurs when viewing a web page to automatically extract useful information.

Large web sites usually use defensive algorithms to protect their data from web scrapers and to limit the number of requests an IP or IP network may send. This has caused an ongoing battle between web site developers and scraping developers.

1.2.3 Report Mining

Report mining is the extraction of data from human readable computer reports. Conventional data extraction requires a connection to a working source system, suitable connectivity standards or an API, and usually complex querying. By using the source system's standard reporting options, and directing the output to a spool file instead of to a printer, static reports can be generated suitable for offline analysis via report mining.This approach can avoid intensive CPU usage during business hours, minimise end user licence costs for ERP customers, and offer very rapid prototyping and development of custom reports. Whereas data scraping and web scraping involve interacting with dynamic output, report mining involves extracting data from files in a human readable format, such as HTML, PDF, or text. These can be easily generated from almost any system by intercepting the data feed to a printer. This approach can provide a quick and simple route to obtaining data without needing to program an API to the source system.

2. Data Curation

Data curation is the organization and integration of data collected from various sources. It involves annotation, publication and presentation of the data such that the value of the data is maintained over time, and the data remain available for reuse and preservation. Data curation includes all the processes needed for principled and controlled data creation, maintenance, and management, together with the capacity to add value to data. In science, data curation may indicate the process of extraction of important information from scientific texts, such as researching articles by experts, and converting it into an electronic format, such as an entry of a biological database.

In the modern era of big data the curation of data has become more prominent, particularly for software processing high volume and complex data systems. The term is also used in historical uses and the humanities, where increasing cultural and scholarly data from digital humanities projects require the expertise and analytical practices of data curation. In broad terms, curation means a range of activities and processes done to create, manage, maintain, and validate a component. Specifically, data curation is the attempt to determine what information is worth saving and for how long.

3. Data Cleansing

Data cleansing, also known as data cleaning or data scrubbing, is the process of altering data in a given storage resource to make sure that it is accurate and correct. There are many ways to pursue data cleansing in various software and data storage architectures; most of them center on the careful review of data sets and the protocols associated with any particular data storage technology.

Data cleansing is sometimes compared to data purging, where old or useless data will be deleted from a data set. Although data cleansing can involve deleting old, incomplete or duplicated data, data cleansing is different from data purging in that data purging usually focuses on clearing space for new data, whereas data cleansing focuses on maximizing the accuracy of data in a system. A data cleansing method may use parsing or other methods to get rid of syntax errors, typographical errors or fragments of records. Careful analysis of a data set can show how merging multiple sets lead to duplication, in which case data cleansing may be used to fix the problem.

Many issues involving data cleansing are similar to problems that archivists, database admin staff and others face around processes like data maintenance, targeted data mining and the extract, transform, load (ETL) methodology, where old data are reloaded into a new data set. These issues often regard the syntax and specific use of command to effect related tasks in database and server technologies like SQL or Oracle. Data cleansing plays a highly important role in many businesses and organizations that rely on large data sets and accurate records for commerce or any other initiative.

New Words

human-readable	['hju:mən'ri:dəbl]	*adj.* 人可读的
accomplished	[ə'kʌmplɪʃt]	*adj.* 技艺高超的；熟练的
interchange	['ɪntətʃeɪndʒ]	*v.* 交换；相互交换
		n. 交换，交替
ambiguity	[ˌæmbɪ'gju:ɪtɪ]	*n.* 含糊，意义不明确
binary	['baɪnərɪ]	*adj.* 二进制的
		n. 二进制数
superfluous	[su:'pɜ:flʊəs]	*adj.* 过多的，多余的；不必要的
commentary	['kɒməntrɪ]	*n.* 评论，评注；注释，注解
irrelevant	[ɪ'reləvənt]	*adj.* 不相干的
unwanted	[ˌʌn'wɒntɪd]	*adj.* 不需要的，无用的；多余的

ad hoc	[ˌæd ˈhɒk]	*adj.* 特别的；临时的；特设的 *adv.* 特别地
inelegant	[ɪnˈelɪgənt]	*adj.* 不优雅的
cope	[kəʊp]	*vi.* 成功地应付，对付
nonsense	[ˈnɒnsns]	*n.* 胡说，废话；荒谬的念头；愚蠢的行为；胡闹 *adj.* 无意义的；荒谬的
variant	[ˈveəriənt]	*n.* 变体；变形，转化 *adj.* 不同的；多样的；变异的；易变的，不定的
collection	[kəˈlekʃn]	*n.* 收集，采集
terminal	[ˈtɜːmɪnl]	*n.* 终端 *adj.* 末端的
auxiliary	[ɔːgˈzɪlɪərɪ]	*adj.* 辅助的；附加的；副的 *n.* 辅助设备
port	[pɔːt]	*n.* 端口，接口
bidirectional	[ˌbaɪdiˈrekʃənl]	*adj.* 双向的
feed	[fiːd]	*vt.* 向……提供 *vi.* 馈送；流入，注入
client	[ˈklaɪənt]	*n.* 客户端
simulate	[ˈsɪmjʊleɪt]	*vt.* 模仿；模拟 *adj.* 模仿的；假装的
automatically	[ˌɔːtəˈmætɪklɪ]	*adv.* 自动地
defensive	[dɪˈfensɪv]	*adj.* 防御用的，防守的
battle	[ˈbætl]	*v.* 与……作战；争斗 *n.* 战争；较量
suitable	[ˈsuːtəbl]	*adj.* 合适的，适当的
intensive	[ɪnˈtensɪv]	*adj.* 加强的，强烈的 *n.* 加强器
prototype	[ˈprəʊtətaɪp]	*n.* 原型
intercept	[ˌɪntəˈsept]	*vt.* 拦截，拦住
curation	[kjʊəˈreɪʃən]	*n.*（对数字信息的）综合处理
principle	[ˈprɪnsɪpl]	*n.* 原则，原理；准则
indicate	[ˈɪndɪkeɪt]	*vt.* 表明，标示，指示
scholarly	[ˈskɒləlɪ]	*adj.* 学术性的；学者的
accurate	[ˈækjərət]	*adj.* 精确的，准确的；正确无误的
incomplete	[ˌɪnkəmˈpliːt]	*adj.* 不完全的；不完备的；未完成的
typographical	[ˌtaɪpəˈgræfɪkl]	*adj.* 印刷上的
fragment	[ˈfrægmənt]	*n.* 碎片；片段，未完成的部分

archivist [ˈɑːkɪvɪst] *n.* 档案保管员
reload [ˌriːˈləʊd] *vt.* 再装，重新加载

Phrases

data scraping 数据抓取
binary data 二进制数
advertisement revenue 广告收入
last resort 最后依靠，最后的补救办法
screen scraping 屏幕抓取
terminal's screen 终端屏幕
output port 输出端口
text form 文本格式
web page 网页
tool kit 工具箱
web site 网站
web server 网络服务器
computer vision 计算机视觉
natural language processing 自然语言处理
spool file 假脱机文件；临时文件
data curation 数据管理
electronic format 电子格式
digital humanities 数字人文
attempt to 试图，尝试
data scrubbing 数据清洗
get rid of 除掉，去掉
syntax error 句法错误

Abbreviations

API（Application Programming Interface） 应用程序接口
XHTML（eXtensible HyperText Markup Language） 可扩展超文本标记语言
JSON（JavaScript Object Notation） JavaScript 对象简谱
DOM（Document Object Model） 文档对象模型
IP（Internet Protocol） 互联网协议
PDF（Portable Document Format） 便携式文档格式

Exercises

【Ex.5】Answer the following questions according to the text.

1. What is data scraping?
2. How is data transfer between programs accomplished normally?
3. What does data scraping often involve?
4. What was screen scraping referred to originally? How was this done?
5. What is a web scraper? What do newer forms of web scraping involve?
6. What do large web sites usually use defensive algorithms to do? What has this caused?
7. What do data scraping and web scraping involve?
8. What is data curation? What does it involve?
9. What is data cleansing?
10. How is data cleansing different from data purging?

参考译文

数据处理（1）

1. 数据预处理

数据预处理是数据挖掘过程中的重要一步。短语“垃圾入，垃圾出”特别适用于数据挖掘和机器学习项目。数据收集方法通常是松散控制的，导致出现超出范围的值（例如，收入：-100）、不可能的数据组合（例如，性别：男性。怀孕：是）、缺失值等。对此类问题的数据未经过仔细筛选就分析可能会产生误导性结果。因此，在进行分析之前，最重要的是数据的表示和质量。通常，数据预处理是机器学习项目中最重要的阶段，特别是在计算生物学中。

如果存在许多不相关和冗余的信息或者噪声和不可靠的数据，则在训练阶段的知识发现更加困难。数据准备和过滤步骤可能需要相当长的处理时间。数据预处理包括清理、实例选择、规范化、转换、特征提取和选择等。数据预处理的产品是最终的训练集。

2. 数据编辑

数据编辑被定义为对收集的调查数据进行审查和调整的过程。目的是控制收集数据的质量。可以手动、在计算机的帮助下或两者相结合来执行数据编辑。数据编辑方法主要包括以下几种。

2.1 交互式编辑

术语交互式编辑通常用于现代计算机辅助手动编辑。国家统计局（NSI）应用的大多数交互式数据编辑工具允许用户在数据输入期间或之后检查特定的编辑，并在必要时立即纠正错误数据。可以遵循以下几种方法来纠正错误数据：

- 重新联系回答者
- 将回答者的数据与其上一年的数据进行比较
- 将回答者的数据与来自类似回答者的数据进行比较
- 使用人类编辑的主题知识

交互式编辑是编辑数据的标准方法，它可用于编辑分类和连续的数据。交互式编辑缩短了完成审核和调整周期过程所需的时间。

2.2 选择性编辑

选择性编辑是识别有影响的错误和异常值的几种方法的总称。选择性编辑技术旨在将交互式编辑应用于精心选择的记录子集，以便将可用于交互式编辑的有限时间和资源分配给那些对发布数字的最终估计质量影响最大的记录。在选择性编辑中，数据分为两个流：

- 关键流
- 非关键流

关键流由更有可能包含有影响的错误的记录组成。这些关键记录以传统的交互方式进行编辑。在非关键流中，不太可能包含有影响的错误的记录。这些记录不以计算机辅助方式编辑。

2.3 宏编辑

宏编辑有两种方法：

- 聚合方法

在发布之前，几乎每个统计机构都遵循这种方法：验证要公布的数字是否合理。这通过将发布表中的数量与先前发布的相同数量进行比较来实现。如果观察到异常值，则对导致可疑数量的各个记录和字段应用宏编辑程序。

- 分布方法

可用数据用于表征变量的分布。然后将所有单个值与分布进行比较。包含可能被视为不常见的值（给定分布）的记录是进一步检查和可能编辑的候选者。

2.4 自动编辑

在自动编辑中，记录由计算机编辑而无须人为干预。可以把单个变量或变量组合值的先验知识制定成一组编辑规则，用于指定或约束允许值。

3. 数据精简

数据精简是将根据经验或实验得出的数字或字母数字信息转换为校正的、有序的和简化的形式。基本概念是将大量数据减少到有意义的部分。

当信息来自仪器读数时，也可能存在从模拟到数字形式的转换。当数据已经是数字形式时，数据的“精简”通常涉及一些编辑、规模调整、编码、排序、整理和生成表格摘要。当观察到是离散的但是潜在的现象是连续的时候，通常需要平滑和插值。通常，在出现读数或测量误差时，要进行数据精简。在确定最可能的值之前，需要对这些误差的性质有所了解。

这些是用于数据精减的常用技术：

- 按规模的某些方面排序。
- 表对角化，从而重新排列表的行和列以使模式更易于查看。
- 大幅度地舍入一个或最多两个有效数字。
- 使用平均值提供视觉焦点和摘要。
- 使用布局和标签来引导眼球。
- 删除图表废物，例如图片和线条。
- 提供简短的口头总结。

4．数据整理

数据整理（有时称为数据调整）是将数据从一种“原始”数据形式转换和映射到另一种格式的过程，目的是使其更适合于各种下游目的（如分析）。数据整理者是执行这些转换操作的人。

这可能包括进一步调整、数据可视化、数据聚合、训练统计模型以及许多其他潜在用法。数据调整过程通常有一系列步骤：开始从数据源以原始形式提取数据，再使用算法（例如排序）“整理”原始数据或将数据解析成预定义的数据结构，最后将生成的内容存入数据接收器以供存储和将来使用。

4.1 典型用途

数据转换通常应用于数据集内的不同实体（例如，字段、行、列、数据值等），并且可以包括诸如提取、解析、加入、标准化、扩充、清理、合并和过滤操作。期望整理后的数据可供下游使用。

接收整理结果数据的可以是个人，例如，将进一步调查数据的数据架构师或数据科学家、将直接在报告中使用数据的业务用户，或者进一步处理数据并将其写入目标（如数据仓库、数据湖或下游应用程序）的系统。

4.2 工作方法

根据输入数据的数量和格式，传统上数据整理是手动执行的（如通过 Excel 等电子表格）或通过 Python 或 SQL 等语言编写的手写脚本。R 语言是一种常用于数据挖掘和统计数据分析的语言，现在也经常用于数据整理。这些流程的其他术语包括数据特许权、数据准备和数据调整。

Unit 9

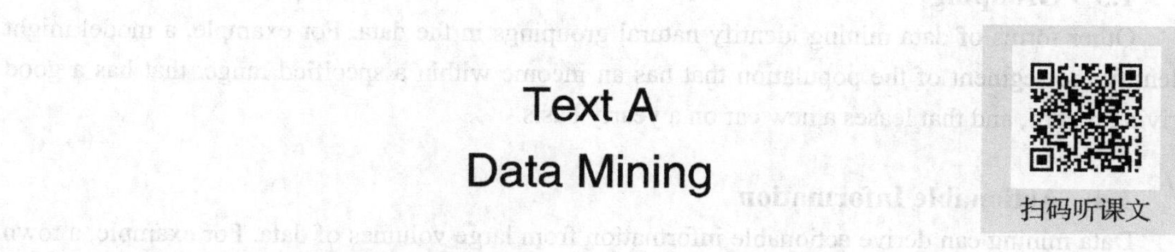

Text A
Data Mining

1. What Is Data Mining

Data mining is the practice of automatically searching large stores of data to discover patterns and trends that go beyond simple analysis. Data mining uses sophisticated mathematical algorithms to segment the data and evaluate the probability of future events. Data mining is also known as Knowledge Discovery in Data (KDD).

The key properties of data mining are:

- Automatic discovery of patterns.
- Prediction of likely outcomes.
- Creation of actionable information.
- Focus on large data sets and databases.
- Answering questions that cannot be addressed through simple query and reporting techniques.

1.1 Automatic Discovery

Data mining is accomplished by building models. A model uses an algorithm to act on a set of data. The notion of automatic discovery refers to the execution of data mining models.

Data mining models can be used to mine the data on which they are built, but most types of models are generalizable to new data. The process of applying a model to new data is known as scoring.

1.2 Prediction

Many forms of data mining are predictive. For example, a model might predict income based on education and other demographic factors. Predictions have an associated probability (How likely is this prediction to be true?). Prediction probabilities are also known as confidence (How confident can

I be of this prediction?).

Some forms of predictive data mining generate rules, which are conditions that imply a given outcome. For example, a rule might specify that a person who has a bachelor's degree and lives in a certain neighborhood is likely to have an income greater than the regional average. Rules have an associated support (What percentage of the population satisfies the rule?).

1.3 Grouping

Other forms of data mining identify natural groupings in the data. For example, a model might identify the segment of the population that has an income within a specified range, that has a good driving record, and that leases a new car on a yearly basis.

1.4 Actionable Information

Data mining can derive actionable information from large volumes of data. For example, a town planner might use a model that predicts income based on demographics to develop a plan for low-income housing. A car leasing agency might use a model that identifies customer segments to design a promotion targeting high-value customers.

1.5 Data Mining and Statistics

There is a great deal of overlap between data mining and statistics. In fact, most of the techniques used in data mining can be placed in a statistical framework. However, data mining techniques are not the same as traditional statistical techniques.

Traditional statistical methods, in general, require a great deal of user interaction in order to validate the correctness of a model. As a result, statistical methods can be difficult to automate. Moreover, they typically do not scale well to very large data sets. They rely on testing hypotheses or finding correlations based on smaller, representative samples of a larger population.

Data mining methods are suitable for large data sets and can be more readily automated. In fact, data mining algorithms often require large data sets for the creation of quality models.

1.6 Data Mining and OLAP

On-Line Analytical Processing (OLAP) can be defined as fast analysis of shared multidimensional data. OLAP and data mining are different but complementary activities.

OLAP supports activities such as data summarization, cost allocation, time series analysis, and what-if analysis. However, most OLAP systems do not have inductive inference capabilities beyond the support for time-series forecast. Inductive inference, the process of reaching a general conclusion from specific examples, is a characteristic of data mining. Inductive inference is also known as computational learning.

OLAP systems provide a multidimensional view of the data, including full support for hierarchies. This view of the data is a natural way to analyze businesses and organizations. Data mining, on the other hand, usually does not have a concept of dimensions and hierarchies.

Data mining and OLAP can be integrated in a number of ways. For example, data mining can be used to select the dimensions for a cube, create new values for a dimension, or create new measures for a cube. OLAP can be used to analyze data mining results at different levels of granularity.

Data Mining can help you construct more interesting and useful cubes. For example, the results of predictive data mining could be added as custom measures to a cube. Such measures might provide information such as "likely to default" or "likely to buy" for each customer. OLAP processing could then aggregate and summarize the probabilities.

1.7 Data Mining and Data Warehousing

Data can be mined whether it is stored in flat files, spreadsheets, database tables, or some other storage format. The important criterion for the data is not the storage format, but its applicability to the problem to be solved.

Proper data cleansing and preparation are very important for data mining, and a data warehouse can facilitate these activities. However, a data warehouse will be of no use if it does not contain the data you need to solve your problem.

Oracle Data Mining requires that the data be presented as a case table in single-record case format. All the data for each record (case) must be contained within a row. Most typically, the case table is a view that presents the data in the required format for mining.

2. What Can Data Mining Do and Not Do

Data mining is a powerful tool that can help you find patterns and relationships within your data. But data mining does not work by itself. It does not eliminate the need to know your business, to understand your data, or to understand analytical methods. Data mining discovers hidden information in your data, but it cannot tell you the value of the information to your organization.

You might already be aware of important patterns as a result of working with your data over time. Data mining can confirm or qualify such empirical observations in addition to finding new patterns that may not be immediately discernible through simple observation.

It is important to remember that the predictive relationships discovered through data mining are not necessarily causes of an action or behavior. For example, data mining might determine that males with incomes between $50,000 and $65,000 who subscribe to certain magazines are likely to buy a given product. You can use this information to help you develop a marketing strategy. However, you should not assume that the population identified through data mining will buy the product just because they belong to this population group.

Data mining does not automatically discover solutions without guidance. The patterns you find through data mining will be very different depending on how you formulate the problem.

To obtain meaningful results, you must learn how to ask the right questions. For example, rather than trying to learn how to improve the response to a direct mail solicitation, you might try to find the characteristics of people who have responded to your solicitations in the past.

To ensure meaningful data mining results, you must understand your data. Data mining algorithms are often sensitive to specific characteristics of the data: Outliers (data values that are very different from the typical values in your database), irrelevant columns, columns that vary together (such as age and date of birth), data coding, and data that you choose to include or exclude. Oracle Data Mining can automatically perform much of the data preparation required by the algorithm. But some of the data preparation is typically specific to the domain or the data mining problem. At any rate, you need to understand the data that was used to build the model in order to properly interpret the results when the model is applied.

3. The Data Mining Process

Figure 9-1 illustrates the phases and the iterative nature of a data mining project. The process flow shows that a data mining project does not stop when a particular solution is deployed. The results of data mining trigger new business questions, which in turn can be used to develop more focused models.

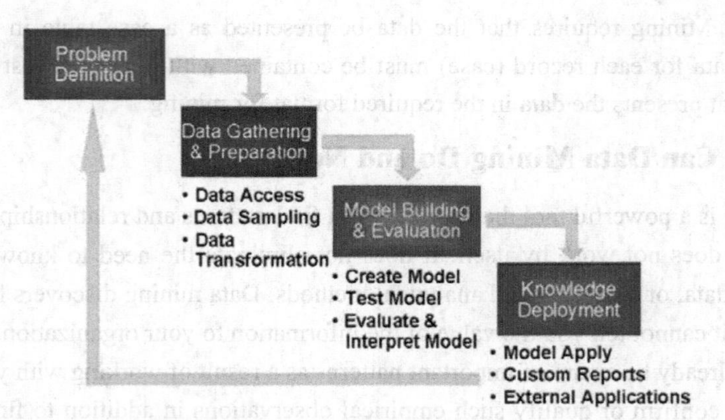

Figure 9-1 The Data Mining Process

3.1 Problem Definition

This initial phase of a data mining project focuses on understanding the project objectives and requirements. Once you have specified the project from a business perspective, you can formulate it as a data mining problem and develop a preliminary implementation plan.

For example, your business problem might be: "How can I sell more of my product to customers?" You might translate this into a data mining problem such as: "Which customers are most likely to purchase the product?" A model that predicts who are most likely to purchase the product must be built on data that describes the customers who have purchased the product in the past. Before building the model, you must assemble the data that is likely to contain relationships between customers who have purchased the product and customers who have not purchased the product. Customer attributes might include age, number of children, years of residence, owners/renters, and so on.

3.2 Data Gathering and Preparation

The data understanding phase involves data collection and exploration. As you take a closer look at the data, you can determine how well it addresses the business problem. You might decide to remove some of the data or add additional data. This is also the time to identify data quality problems and to scan for patterns in the data.

The data preparation phase covers all the tasks involved in creating the case table you will use to build the model. Data preparation tasks are likely to be performed multiple times, and not in any prescribed order. Tasks include table, case, and attribute selection as well as data cleansing and transformation. For example, you might transform a DATE_OF_BIRTH column to AGE; you might insert the average income in cases where the INCOME column is null.

Thoughtful data preparation can significantly improve the information that can be discovered through data mining.

3.3 Model Building and Evaluation

In this phase, you select and apply various modeling techniques and calibrate the parameters to optimal values. If the algorithm requires data transformations, you will need to step back to the previous phase to implement them.

In preliminary model building, it often makes sense to work with a reduced set of data (fewer rows in the case table), since the final case table might contain thousands or millions of cases.

At this stage of the project, it is time to evaluate how well the model satisfies the originally-stated business goal.

3.4 Knowledge Deployment

Knowledge deployment is the use of data mining within a target environment. In the deployment phase, insight and actionable information can be derived from data.

Deployment can involve scoring (the application of models to new data), the extraction of model details (for example, the rules of a decision tree), or the integration of data mining models within applications, data warehouse infrastructure, or query and reporting tools.

Because Oracle Data Mining builds and applies data mining models inside Oracle Database, the results are immediately available. BI reporting tools and dashboards can easily display the results of data mining. Additionally, Oracle Data Mining supports scoring in real time: Data can be mined and the results can be returned within a single database transaction. For example, a sales representative could run a model that predicts the likelihood of fraud within the context of an online sales transaction.

New Words

practice	[ˈpræktɪs]	*n.* 练习；实践 *vi.* 实行；练习 *vt.* 实行，实践
evaluate	[ɪˈvæljʊeɪt]	*v.* 评价，估价
probability	[ˌprɒbəˈbɪlɪtɪ]	*n.* 可能性；概率
event	[ɪˈvent]	*n.* 事件，大事；活动，经历；结果
outcome	[ˈaʊtkʌm]	*n.* 结果，成果
actionable	[ˈækʃənəbl]	*adj.* 可行动性；可执行的
generalizable	[ˈdʒenərəlaɪzəbl]	*adj.* 可概括的，可归纳的
scoring	[ˈskɔːrɪŋ]	*n.* 得分
demographic	[ˌdeməˈgræfɪk]	*adj.* 人口统计学的
factor	[ˈfæktə]	*n.* 因素，要素
confidence	[ˈkɒnfɪdəns]	*n.* 置信度
neighborhood	[ˈneɪbəhʊd]	*n.* 地区
regional	[ˈriːdʒənl]	*adj.* 地区的，区域的
population	[ˌpɒpjʊˈleɪʃn]	*n.* 人口
promotion	[prəˈməʊʃn]	*n.* 促进，增进；提升，升级；（商品等的）推广
overlap	[ˌəʊvəˈlæp]	*n.* 重叠部分 *v.* 重叠
correctness	[kəˈrektnɪs]	*n.* 正确性
hypothesis	[haɪˈpɒθɪsɪs]	*n.* 假设，假说
sample	[ˈsɑːmpl]	*n.* 样本，样品 *vt.* 取……的样品
summarization	[ˌsʌmərɪˈzeɪʃən]	*n.* 摘要，概要；概括
inductive	[ɪnˈdʌktɪv]	*adj.* 归纳的；归纳法的
inference	[ˈɪnfərəns]	*n.* 推理；推断；推论
conclusion	[kənˈkluːʒn]	*n.* 结论；断定，决定；推论
cube	[kjuːb]	*n.* 立方形，立方体；立方，三次幂 *vt.* 求……的立方
construct	[kənˈstrʌkt]	*vt.* 构建，建造；构成；创立
spreadsheet	[ˈspredʃiːt]	*n.* 电子表格
relationship	[rɪˈleɪʃnʃɪp]	*n.* 关系；联系
hide	[haɪd]	*vt.* 隐藏，隐匿
empirical	[ɪmˈpɪrɪkl]	*adj.* 凭经验的；以观察或实验为依据的

discernible	[dɪ'sɜːnəbl]	*adj.* 可识别的；可辨别的
action	['ækʃn]	*n.* 行动，活动；功能，作用；手段
behavior	[bɪ'heɪvɪə]	*n.* 行为；态度
solution	[sə'luːʃn]	*n.* 解决方案，答案
formulate	['fɔːmjʊleɪt]	*vt.* 构想出，规划；确切地阐述；用公式表示
obtain	[əb'teɪn]	*vt.* 获得，得到
solicitation	[ˌsəlɪsɪ'teɪʃn]	*n.* 请求，征求；询价
exclude	[ɪk'skluːd]	*vt.* 排除，不包括
illustrate	['ɪləstreɪt]	*vt.*（用示例、图画等）说明；给……加插图
trigger	['trɪgə]	*vt.* 引发，触发
requirement	[rɪ'kwaɪəmənt]	*n.* 需求，要求
preliminary	[prɪ'lɪmɪnərɪ]	*adj.* 初步的，初级的；预备的；开端的 *n.* 准备工作；初步措施
plan	[plæn]	*n.* 计划，打算 *v.* 规划，计划，打算
assemble	[ə'sembl]	*v.* 集合，收集
exploration	[ˌeksplə'reɪʃn]	*n.* 探测；搜索，研究
scan	[skæn]	*v.* 审视
prescribe	[prɪ'skraɪb]	*vt.* 指定，规定 *vi.* 建立规定，法律或指示
insert	[ɪn'sɜːt]	*vt.* 插入
evaluation	[ɪˌvæljʊ'eɪʃn]	*n.* 评估；估价
calibrate	['kælɪbreɪt]	*vt.* 校准；使标准化，使合标准
optimal	['ɒptɪməl]	*adj.* 最佳的，最优的
insight	['ɪnsaɪt]	*n.* 洞察力，洞悉；直觉，眼光；领悟
dashboard	['dæʃbɔːd]	*n.* 仪表板，仪表盘
likelihood	['laɪklihʊd]	*n.* 可能，可能性
fraud	[frɔːd]	*n.* 欺诈；骗子；伪劣品，冒牌货

Phrases

automatic discovery	自动发现
data mining model	数据挖掘模型
high-value customer	高价值客户
in order to ...	为了……
statistical method	统计方法

rely on	依赖，依靠；信任
data mining algorithm	数据挖掘算法
be defined as	被定义为
multidimensional data	多维数据
cost allocation	成本分摊
time series analysis	时间序列分析，时序分析
computational learning	计算学习
be integrated in ...	被集成到……中
flat file	平面文件
be aware of	知道；意识到
marketing strategy	市场战略，营销战略
direct mail	直接邮件，直接邮寄广告
at any rate	无论如何，至少
focus on	聚焦于
implementation plan	实施计划，实施方案
translate ... into ...	把……转换为……，把……翻译为……
optimal value	最佳值，最优值
knowledge deployment	知识部署
decision tree	决策树

Abbreviation

KDD（Knowledge Discovery in Database） 数据知识发现

Exercises

【Ex.1】Answer the following questions according to the text.

1. What is data mining?
2. What are the key properties of data mining?
3. What can data mining models be used to do? What is known as scoring?
4. What do traditional statistical methods require, in general, in order to validate the correctness of a model?
5. What can On-Line Analytical Processing (OLAP) be defined as? What does it support?
6. What is the important criterion for the data?
7. What is data mining? What does it do?
8. What are data mining algorithms often sensitive to?

9. What does this initial phase of a data mining project focus on? What can you do once you have specified the project from a business perspective?

10. What is knowledge deployment? What can it involve?

【Ex.2】Translate the following terms or phrases from English into Chinese and vice versa.

1. data mining algorithm	1. ________
2. decision tree	2. ________
3. knowledge deployment	3. ________
4. optimal value	4. ________
5. statistical method	5. ________
6. 时间序列分析，时序分析	6. ________
7. *vt.*引发，触发	7. ________
8. *n.*样本，样品 *vt.*取……的样品	8. ________
9. *n.*可能性；概率	9. ________
10. *n.*洞察力，洞悉；领悟	10. ________

【Ex.3】Translate the following passages into Chinese.

Decision Tree

A decision tree is a diagram or chart that people use to determine a course of action or show a statistical probability. Each branch of the decision tree represents a possible decision, outcome, or reaction. The farthest branches of the tree represent the end results.

A decision tree is a graphical depiction of a decision and every potential outcome or result of making that decision. By displaying a sequence of steps, decision trees give people an effective and easy way to visualize and understand the potential options of a decision and its range of possible outcomes. The decision tree also helps people identify every potential option and weigh each course of action against the risks and rewards each option can yield.

An organization may deploy decision trees as a kind of decision support system. The structured model allows the reader of the chart to see how and why one choice may lead to the next, with the use of the branches indicating mutually exclusive options. The structure allows users to take a problem with multiple possible solutions and to display those solutions in a simple, easy-to-understand format that also shows the relationship between different events or decisions.

In the decision tree, each end result has an assigned risk and reward weight or number. If a person uses a decision tree to make a decision, he or she looks at each final outcome and assesses the benefits and drawbacks. The tree itself can span as long or as short as needed in order to come to a proper conclusion.

【Ex.4】Fill in the blanks with the words given below.

attributes	groups	patterns	application	challenge
parameters	mining	algorithms	columns	predicts

Data Mining Algorithms

The data mining algorithm is the mechanism that creates a data mining model. To create a model, an algorithm first analyzes a set of data and looks for specific ___1___ and trends. The algorithm uses the results of this analysis to define the parameters of the mining model. These ___2___ are then applied across the entire data set to extract actionable patterns and detailed statistics.

The mining model that an algorithm creates can take various forms, including:

- A set of rules that describe how products are grouped together in a transaction.
- A decision tree that ___3___ whether a particular customer will buy a product.
- A mathematical model that forecasts sales.
- A set of clusters that describe how the cases in a data set are related.

Microsoft SQL Server Analysis Services provides several algorithms for use in your data ___4___ solutions. These algorithms are a subset of all the algorithms that can be used for data mining. You can also use third-party ___5___ that comply with the OLEDB for Data Mining specification.

Analysis Services includes the following algorithm types:

- Classification algorithms predict one or more discrete variables, based on the other ___6___ in the data set.
- Regression algorithms predict one or more continuous variables, such as profit or loss, based on other attributes in the dataset.
- Segmentation algorithms divide data into ___7___, or clusters, of items that have similar properties.
- Association algorithms find correlations between different attributes in a dataset. The most common ___8___ of this kind of algorithm is for creating association rules, which can be used in a market basket analysis.

Sequence analysis algorithms summarize frequent sequences or episodes in data, such as a Web path flow.

Choosing the best algorithm to use for a specific business task can be a ___9___. While you can use different algorithms to perform the same business task, each algorithm produces a different result, and some algorithms can produce more than one type of result. For example, you can use the Microsoft Decision Trees algorithm not only for prediction, but also as a way to reduce the number of ___10___ in a data set, because the decision tree can identify columns that do not affect the final mining model.

Text B
Data Mining Algorithms

The main tools in a data miner's arsenal are algorithms. Today, I'm going to look at the top 10 data mining algorithms, and make a comparison of how they work and what each can be used for.

Algorithms are a set of instructions that a computer can run. They aren't specific to one programming language and can even be written down in plain English.

In data mining, clever algorithms are used to find patterns in large sets of data, and help classify new information. The applications for these are limitless. Let's take a look at some examples of data mining algorithms.

1. C4.5

The first on this list of data mining algorithms is C4.5. It is a classifier, meaning it takes in data and attempts to guess which class it belongs to.

C4.5 is also a supervised learning algorithm and needs training data. Data scientists run C4.5 on the training data to build a decision tree. This can then be used to classify new information.

Let's take a look at an example. Say we had a sample of data showing who survived the sinking of the Titanic. We also know their sex, age, and the number of siblings they had on board. Running C4.5 might generate a decision tree like the one below (See Figure 9-2).

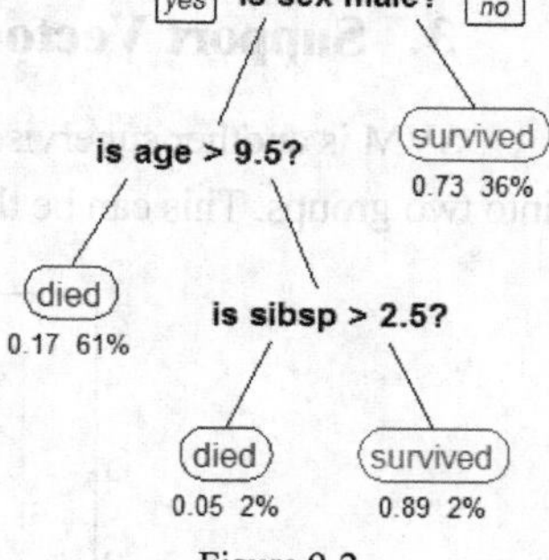

Figure 9-2

Now we can use this decision tree on new data to predict whether a passenger survived or not.

The main problem with C4.5 is over fitting when using noisy data. This means the decision tree it generates pays too much attention to outliers and will make mistakes if there are too many in the data set.

2. K-Means

K-means is very different type of data mining algorithm from C4.5. It is an unsupervised learning algorithm, meaning it doesn't need training data, and works even your data isn't already marked or classified.

Rather than classifying data, the goal of k-means is to group data points together based on how similar they are. These groups are called "Clusters". A very simple example might be to group a set of people together by age and blood pressure. The "K" simply tells us there can be any number of groups.

It's a popular data mining algorithm because it's simple to be implemented and works quite well. Here's an example of a data set that has been grouped using this techinque.

K-means works by finding points called "centroids", then assigns each point in the data to a cluster based on its closest centroid. As you can guess, the trick here is to find the optimum number and position of centroids to group the data properly.

This is done in a very clever way, here's a basic version of the algorithm:

(1) Choose the number K of centroids you want.

(2) Randomly select positions for the centroids in multidimensional space.

(3) Assign each data point to its closest centroid, creating K clusters.

(4) Readjust the positions of the centroids to be the average (or mean) position of these new clusters (this is why it's called K-"means").

(5) Repeat steps 3 and 4 until the centroids no longer move!

Once the centroids stop moving, the algorithm has finished and the data has been clustered. For best results, k-means is usually run multiple times with different random starting points. The one with the tightest clusters is probably the best result.

K-medoids is a similar version of this algorithm but is more resilient to outliers.

3. Support Vector Machines (SVM)

SVM is another supervised classifier algorithm like C4.5. The difference is that it only classifies data into two groups. This can be thought of as a line separating data points on a graph (See Figure 9-3).

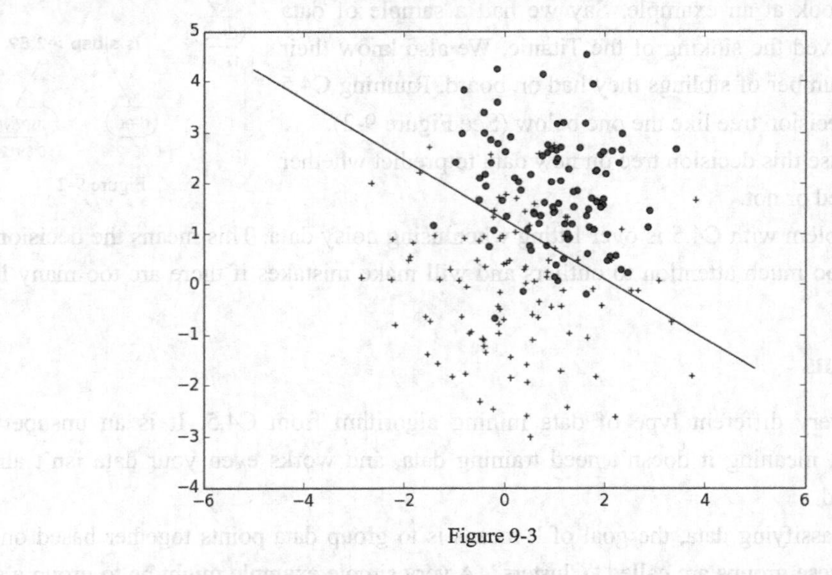

Figure 9-3

The thing that makes the Support Vector Machine algorithm really cool is that it can find complex curved lines separating points more accurately than a straight line. It does this with a clever trick using dimensions.

SVM takes a data set and projects the points into a higher dimension. By doing so, it can separate the points into two groups more easily. Here's a great example (See Figure 9-4).

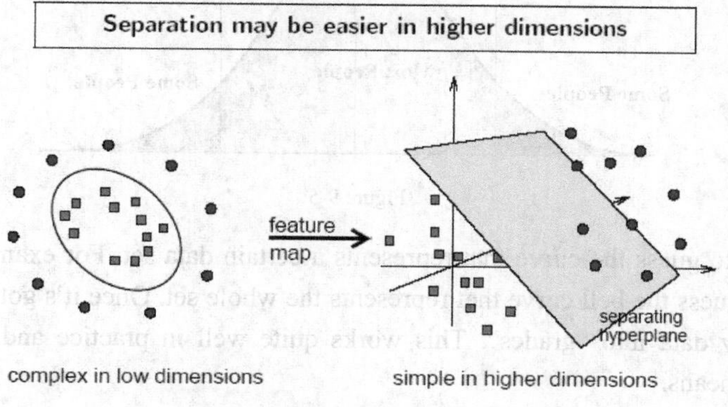

Figure 9-4

The two-dimensional data on the left can't be separated with a straight line. To fix this, SVM projects points into the 3-dimensional space shown on the right. Now, the data can easily be split into two classes with a plane that SVM can find automatically.

Of course, this won't always work, so you'll need to experiment with different classifier algorithms to see which one is best for a particular problem.

4. Apriori

Apriori is an algorithm that learns which items in a data are commonly associated with each other. This can be very useful for grouping similar items together in tables.

For example, your company might have a database of customers and transactions. By running the Apriori algorithm on this data, you might find that coffee beans are commonly bought together with coffee machines. You could use this information to display product recommendations at the right time to increase sales.

The algorithm itself is quite complicated. It is usually considered unsupervised learning, as it can be used to explore sets of unlabelled data. But, it can be modified to give classifications, too.

5. Expectation-Maximization (EM)

EM is a clustering algorithm that uses statistical models to group data.

You might know that the results of test scores usually follow a bell curve. That is, most of the test results will fall somewhere in the middle with few getting very high scores and few getting very low scores. Bell curves are used all the time in statistics and show up in all kinds of data. On paper it looks something like this (See Figure 9-5).

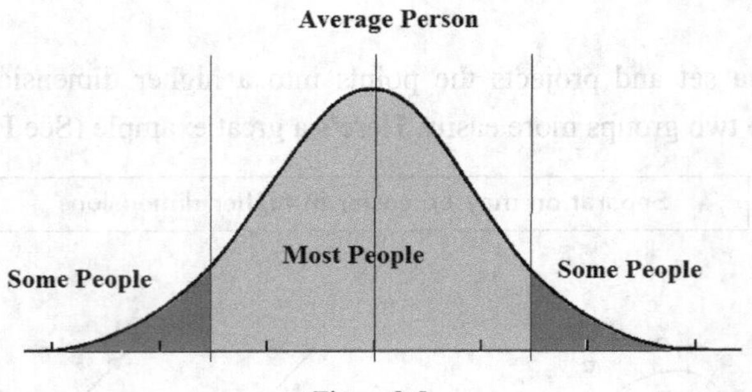

Figure 9-5

EM attempts to guess the curve that represents a certain data set. For example, given a sample set of test scores, guess the bell curve that represents the whole set. Once it's got this curve, it can be used to group new data into "grades". This works quite well in practice and has many different applications to K-means.

6. PageRank

PageRank is one of my favorite algorithms. It actually affects your life every day, and mine. You probably know it better as it is the main algorithm that powers Google searches.

Back before PageRank was invented by the guys at Google, internet searches were terrible. They basically counted the number of times a keyword appeared on a web page to determine how relevant it was to that keyword. This led to lots of spam and bad results. So how do you fix that?

You can think of PageRank as a kind of voting algorithm. It counts the number of times a web page is linked to by other pages. The more links pointing to a page, the more important that web page is considered. Also, if a web page is found to be important, its links will also be more important, and carry more weight.

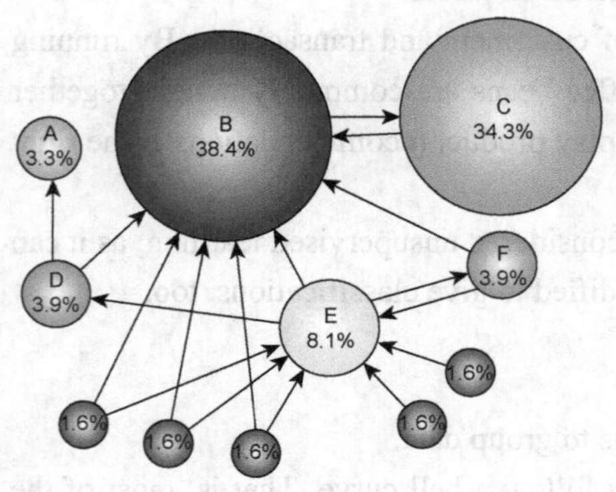

Figure 9-6

PageRank can be used for more than just ranking web pages. It can be extended to any kind of graph where things are linked together. Here's an example of a linked graph that has been run through page rank. The sizes and percentages show how "important" the nodes are found to be based on links to them (See Figure 9-6).

7. AdaBoost

AdaBoost is a data mining algorithm that builds a good classifier out of lots of bad classifiers.

In machine learning, a bad learner is a classifier that doesn't work very well, barely better than random chance. However, if you have a few of them, you might be able to combine them in a way that ends up being pretty good!

AdaBoost does this by using a training set of data, so it is supervised learning. It iterates through the training data on each bad learner, working out which is the best at each step. Each new round, the misclassified points are more important. Eventually, a more complex decision tree is built from the simpler parts. This new tree should be better at classifying new data than any of the original learning algorithms.

The cool thing about AdaBoost is that it can be used with other algorithms. The 'bad learner' building blocks can be the simple decision trees built by other algorithms.

8. K-Nearest Neighbours (KNN)

K-nearest-neighbours is one of the more simple data mining algorithms. Like most of the list, it is also a classifier (these are very important in data mining). However, it does things a little differently from the ones I've already described.

K-means, AdaBoost, and SVM are all eager learners. They take training data and build a classifier that can be used on new data.

KNN is a lazy learner. It doesn't actually do anything until it is given new, unlabeled data. When this happens, it finds the 'k' nearest neighbors to the new data point (i.e. the ones that are most similar) and assigns a class similar to those. Here's an example (See Figure 9-7).

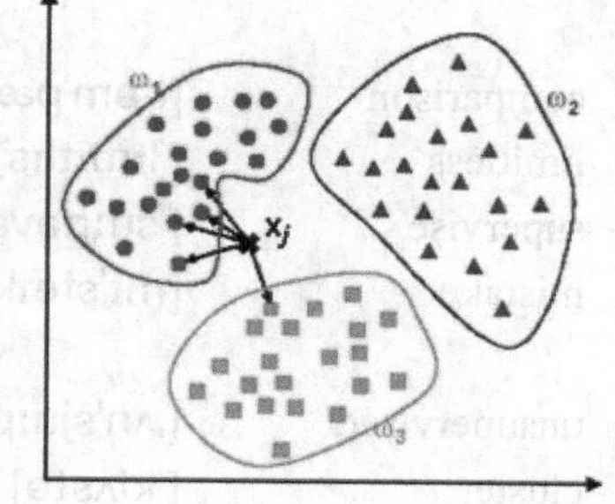

Figure 9-7

The new point X is given as new input. Of its five closest neighbors, four of them are in the red class and one is green. Based on this, X is classified as red, as most of its neighbors are red. Pretty simple!

9. Naive Bayes

This actually refers to a group of algorithms that share a common-naive-assumption.

The assumption is that each feature in a set of data is completely uncorrelated with all other features in the set. For example, say we have data about hospital patients. The features of the data include zip code, height, and age. Those three things probably aren't correlated-your zip code doesn't determine your height or age very much. Here the Naive Bayes assumption is fine.

Using this assumption, some very simple math can give us a classifier that can predict the class of new data.

But, this assumption isn't always true. Let's say we extend our hospital patient data set to include things like blood pressure, cholesterol, and weight. These probably are correlated, so will the algorithm still work?

The answer is usually yes. Naive Bayes classifiers work surprisingly well, even on data with correlated features.

10. CART

The last data mining algorithm on the list is CART, or Classification And Regression Trees. It's an algorithm used to build decision trees, just like many of the other algorithms we've discussed.

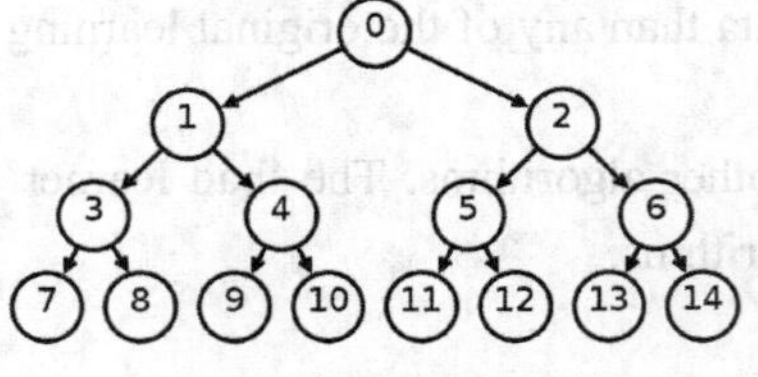

Figure 9-8

CART can be thought of as a more statistically grounded version of C4.5, and can lead to more robust decision trees. One of its defining features is that it constructs binary trees, meaning each internal node has exactly two edges (See Figure 9-8).

The reason to use this algorithm over C4.5 is that it is less susceptible to outliers. That means it will work better on a noisy data set where not all data points are correctly classified.

New Words

comparison	[kəm'pærɪsn]	*n.* 比较，对照
limitless	['lɪmɪtlɪs]	*adj.* 无限制的，无界限的
supervise	['suːpəvaɪz]	*v.* 监督；管理；指导
mistake	[mɪ'steɪk]	*n.* 错误，过失，误解；差错；失策，失误 *v.* 弄错，误解
unsupervised	[ˌʌn'sjuːpəvaɪzd]	*v.* 无人监督，无人管理
cluster	['klʌstə]	*n.* 丛；簇，串；群 *vi.* 丛生；群聚 *vt.* 使密集，使聚集
centroid	['sentrɔɪd]	*n.* 质心，矩心
randomly	['rændəmlɪ]	*adv.* 随机地，随便地
readjust	[ˌriːə'dʒʌst]	*v.* 再整理，再调整
separate	['sepəreɪt]	*v.* 分开；（使）分离；区分；隔开
graph	[grɑːf]	*n.* 图表，曲线图 *vt.* 用曲线图表示，把……绘入图表
recommendation	[ˌrekəmen'deɪʃn]	*n.* 推荐；建议
unlabelled	[ʌn'leɪbld]	*adj.* 未标记的；无标号的
classification	[ˌklæsɪfɪ'keɪʃn]	*n.* 分类；分级；类别
terrible	['terəbl]	*adj.* 可怕的；劣质的；极度的
keyword	['kiːwɜːd]	*n.* 关键字，关键词
spam	[spæm]	*n.* 垃圾邮件 *vt.* 向……发送垃圾函件

voting	[ˈvəʊtɪŋ]	*n.* 表决，选举，投票
link	[lɪŋk]	*n.* 链接；关联 *v.* 链接
rank	[ræŋk]	*v.* 排列；把……分类 *n.* 等级；次序，顺序；行列
percentage	[pəˈsentɪdʒ]	*n.* 百分比，百分率；比例
combine	[kəmˈbaɪn]	*v.* 组合，联合；使结合
misclassify	[mɪsˈklæsɪfaɪ]	*vt.* 对……进行错误的分类
lazy	[ˈleɪzi]	*adj.* 懒惰的；没精打采的；慢吞吞的
naive	[naɪˈiːv]	*adj.* 单纯的
assumption	[əˈsʌmpʃn]	*n.* 假定，假设
uncorrelated	[ʌnˈkɒrɪleɪtɪd]	*adj.* 无关联的
cholesterol	[kəˈlestərɒl]	*n.* 胆固醇
surprisingly	[səˈpraɪzɪŋlɪ]	*adv.* 惊人地，出人意外地
regression	[rɪˈgreʃn]	*n.* 回归
discuss	[dɪˈskʌs]	*vt.* 讨论，谈论；论述，详述；商量
exactly	[ɪgˈzæktlɪ]	*adv.* 精确地；确切地

Phrases

supervised learning algorithm	监督学习算法
unsupervised learning algorithm	无监督学习算法
training data	训练数据
blood pressure	血压
multidimensional space	多维空间
straight line	直线
bell curve	钟形曲线
clustering algorithm	类聚算法
a kind of ...	……的一种
voting algorithm	投票算法
be linked to	链接到
page rank	网页排名
regression tree	回归树
internal node	内部节点
noisy data set	含噪数据集

Abbreviations

SVM（Support Vector Machines） 支持向量机
EM（Expectation-Maximization） 期望最大化
KNN（K-Nearest Neighbours） K 最近邻
CART（Classification and Regression Trees） 分类和回归树

Exercises

【Ex.5】Answer the following questions according to the text.

1. What is the first on this list of data mining algorithms? What is it?
2. Is K-means the same type of data mining algorithm as C4.5? What is it? What is its goal?
3. What is the thing that makes the Support Vector Machine algorithm really cool?
4. What is Apriori? What can it be very useful for?
5. What is EM? What does it attempt to do?
6. What can you think of PageRank as? What does it do?
7. What is AdaBoost? What is a bad learner in machine learning?
8. Is KNN a hardworking learner? What does it do?
9. What does Naive Bayes refer to? What is the assumption?
10. What can CART be thought of as? What is one of its defining features?

参考译文

数据挖掘

1. 什么是数据挖掘

数据挖掘是自动搜索大型数据存储区以发现那些靠简单分析无法实现的模式和趋势。数据挖掘使用复杂的数学算法来划分数据并评估未来事件的概率。数据挖掘也称为数据中的知识发现（KDD）。

数据挖掘的关键特性是：

- 自动发现模式。
- 预测可能的结果。
- 创建可操作的信息。
- 专注于大型数据集和数据库。
- 数据挖掘可以回答通过简单查询和报告技术无法解决的问题。

1.1 自动发现

数据挖掘是通过构建模型来完成的。模型使用算法来处理一组数据。自动发现的概念是指数据挖掘模型的执行。

数据挖掘模型可用于挖掘构建它们的数据，但大多数类型的模型可推广到新数据。将模型应用于新数据的过程称为评分。

1.2 预测

多种形式的数据挖掘都是预测性的。例如，模型可能会根据教育和其他人口统计因素预测收入。预测具有相关概率（这种预测有多大可能是真实的?），预测概率也称为置信度（我对这种预测有多大信心?）。

某些形式的预测数据挖掘会生成规则，这些规则是表明给定结果的条件。例如，规则可能规定拥有学士学位并居住在某个社区的人的收入可能高于该地区的平均水平，规则有相关的支持（人口中有多少百分比符合这些规则?）。

1.3 分组

其他形式的数据挖掘识别数据中的自然分组。例如，模型可以识别收入在指定范围内、具有良好驾驶记录并且每年租赁新车的那个群体。

1.4 可行动的信息

数据挖掘可以从大量数据中获取可行动的信息。例如，城镇规划者可能会使用一种模型来预测基于人口统计的收入，以制定低收入住房计划。汽车租赁代理商可以使用模型来识别细分客户，以设计针对高价值客户的促销。

1.5 数据挖掘和统计

数据挖掘和统计之间存在很多重叠。实际上，数据挖掘中使用的大多数技术都可以放在统计框架中。但是，数据挖掘技术与传统统计技术不同。

通常，传统的统计方法需要大量的用户交互才能验证模型的正确性。结果，统计方法可能难以自动化。此外，它们通常不能很好地扩展到非常大的数据集。它们依赖于检验假设或基于较大的群体的较小的代表性样本找到相关性。

数据挖掘方法适用于大型数据集，并且可以更容易地自动化。实际上，数据挖掘算法通常需要大量数据集来创建质量模型。

1.6 数据挖掘和 OLAP

联机分析处理（OLAP）可以定义为共享多维数据的快速分析，OLAP 和数据挖掘是不同但互补的活动。

OLAP 支持数据汇总、成本分摊、时间序列分析和假设分析等活动。但是，大多数 OLAP 系统除了支持时间序列预测外没有归纳推理功能。归纳推理是从具体实例得出一般结论的过程，是数据挖掘的一个特征。归纳推理也称为计算学习。

OLAP 系统提供数据的多维视图，包括完全支持层次结构。这种数据视图是分析企业和组织的自然方式。而数据挖掘通常没有维度和层次结构的概念。

数据挖掘和 OLAP 可以通过多种方式集成。例如，数据挖掘可用于选择多维数据集的维度，为维度创建新值或为多维数据集创建新度量。OLAP 可用于分析不同粒度级别的数据挖掘结果。

数据挖掘可以帮助构建更有趣和有用的多维数据集。例如，可以将预测数据挖掘的结果作为自定义度量添加到多维数据集。这些措施可能会为每个客户提供诸如“可能违约”或“可能购买”等信息。然后，OLAP 处理可以聚合和汇总概率。

1.7　数据挖掘和数据仓库

无论数据是存储在平面文件、电子表格、数据库表还是一些其他存储格式中，都可以实施挖掘。数据的重要标准不是存储格式，而是它对要解决的问题的适用性。

正确的数据清理和准备对于数据挖掘非常重要，数据仓库可以促进这些活动。但是，如果数据仓库不包含解决问题所需的数据，则它将毫无用处。

Oracle Data Mining 要求将数据显示为单记录格式的案例表。每个记录（案例）的所有数据必须包含在一行中。最典型的情况是，案例表是一个视图，用挖掘所需的格式显示数据。

2. 数据挖掘能做什么和不能做什么

数据挖掘是一个功能强大的工具，可以帮助你查找数据中的模式和关系。但是数据挖掘不能单独工作。它需要了解你的业务、了解你的数据或理解分析方法。数据挖掘可以发现数据中的隐藏信息，但它无法判断信息对组织的价值。

随着时间的推移，你可能已经意识到了重要的模式。除了寻找通过简单观察可能无法立即辨别的新模式之外，数据挖掘还可以确认或限定此类经验观察。

重要的是要记住，通过数据挖掘发现的预测关系不一定是活动或行为的原因。例如，数据挖掘可能会确定订阅某些杂志的收入在 50 000 美元到 65 000 美元之间的男性可能会购买给定的产品。你可以使用此信息来帮助制定营销策略。但是，你不应该假设通过数据挖掘确定的这个群体就会购买该产品，只因为他们属于这一群体。

数据挖掘不会在没有指导的情况下自动发现解决办法。通过数据挖掘找到的模式会依据制定问题的方式而有很大差异。

要获得有意义的结果，必须学习如何提出正确的问题。例如，你可能会尝试找到过去对你的请求做出回应的人的特征，而不是尝试学习如何改进对直接邮件征集的响应。

要确保有意义的数据挖掘结果，必须了解你的数据。数据挖掘算法通常对数据的特定特征敏感：异常值（与数据库中的典型值非常不同的数据值）、不相关的列、不同的列（例如年龄和出生日期）、数据编码，以及你选择包含或排除的数据。Oracle Data Mining 可以自动执行算法所需的大部分数据准备。但是一些数据准备针对特定域或特定的数据挖掘问题。无论如何，你需要了解用于构建模型的数据，以便在应用模型时正确解释结果。

3. 数据挖掘过程

图 9-1 说明了数据挖掘项目的阶段和迭代性质。流程表明，在部署特定解决方案时数据挖掘项目不会停止。数据挖掘的结果触发了新的业务问题，而这些问题又可用于开发更聚焦的模型。

3.1 问题定义

数据挖掘项目的初始阶段侧重于理解项目的目标和要求。从业务角度指定项目后，可以将其设定为数据挖掘问题并制定初步实施计划。

例如，你的业务问题可能是："我如何向客户销售更多产品？"你可能会将此转换为数据挖掘问题，例如："哪些客户最有可能购买该产品？"预测最有可能购买该产品的模型必须建立在描述过去购买过该产品的客户的数据之上。在构建模型之前，你必须汇总可能包含已购买该产品的客户与未购买该产品的客户之间关系的数据。客户属性可能包括年龄、子女人数、居住年限、业主/租房者等。

3.2 数据收集和准备

数据理解阶段涉及数据收集和探究。当仔细查看数据时，可以确定它解决业务问题的程度。你可能决定删除一些数据或添加其他数据。这也正是识别数据质量问题和审视数据模式的时候。

数据准备阶段涵盖了涉及用于构建模型的案例表的所有任务。数据准备任务可能会多次执行，而不是按照任何规定的顺序执行。任务包括表、案例和属性选择以及数据清理和转换。例如，你可以将 DATE_OF_BIRTH 列转换为 AGE；你可以在 INCOME 列为空的情况下插入平均收入。

周到的数据准备可以显著改善通过数据挖掘可以发现的信息。

3.3 模型构建和评估

在此阶段，可以选择并应用各种建模技术，并将参数校准为最佳值。如果算法需要数据转换，则需要返回上一阶段来实现它们。

在初步模型构建中，使用精简的数据集（案例表中的行数较少）通常是有意义的，因为最终案例表可能包含数千或数百万个案例。

在项目的这个阶段，是评估模型满足初期业务目标程度的时候。

3.4 知识部署

知识部署是在目标环境中使用数据挖掘。在部署阶段，可以从数据中获得洞察力和可行动的信息。

部署可以涉及评分（模型应用于新数据）、模型细节的提取（例如决策树的规则），或应用程序、数据仓库基础结构或查询和报告工具中的数据挖掘模型的集成。

由于 Oracle Data Mining 在 Oracle 数据库中构建和应用数据挖掘模型，因此可立即获得结果。BI 报告工具和仪表板可以很容易地显示数据挖掘的结果。此外，Oracle Data Mining 支持实时评分：可以挖掘数据并在单个数据库事务中返回结果。例如，销售代表可以运行一个模型，该模型能预测在线销售交易情景中欺诈的可能性。

Unit 10

Text A
Apache Hadoop

Hadoop is an open-source software framework for storing data and running applications on clusters of commodity hardware. It provides massive storage for any kind of data, enormous processing power and the ability to handle virtually limitless concurrent tasks or jobs.

1. Why Is Hadoop Important

- Ability to store and process huge amounts of any kind of data, quickly. With data volumes and varieties constantly increasing, especially from social media and the Internet of Things (IoT), that's a key consideration.
- Computing power. Hadoop's distributed computing model processes big data fast. The more computing nodes you use, the more processing power you have.
- Fault tolerance. Data and application processing are protected against hardware failure. If a node goes down, jobs are automatically redirected to other nodes to make sure the distributed computing does not fail. Multiple copies of all data are stored automatically.
- Flexibility. Unlike traditional relational databases, you don't have to preprocess data before storing it. You can store as much data as you want and decide how to use it later. That includes unstructured data like text, images and videos.
- Low cost. The open-source framework is free and uses commodity hardware to store large quantities of data.
- Scalability.You can easily grow your system to handle more data simply by adding nodes. Little administration is required.

2. What Are the Challenges of Using Hadoop

MapReduce programming is not a good match for all problems. It's good for simple information requests and problems that can be divided into independent units, but it's not efficient for iterative and interactive analytic tasks. MapReduce is file intensive. Because the nodes don't intercommunicate except through sorts and shuffles, iterative algorithms require multiple map-shuffle/sort-reduce phases to complete. This creates multiple files between MapReduce phases and is inefficient for advanced analytic computing.

There's a widely acknowledged talent gap. It can be difficult to find entry-level programmers who have sufficient Java skills to be productive with MapReduce. That's one reason distribution providers are racing to put relational (SQL) technology on top of Hadoop. It is much easier to find programmers with SQL skills than MapReduce skills. And, Hadoop administration seems part art and part science, requiring low-level knowledge of operating systems, hardware and Hadoop kernel settings.

Data security. Another challenge centers around the fragmented data security issues, though new tools and technologies are surfacing. The Kerberos authentication protocol is a great step toward making Hadoop environments secure.

Full-fledged data management and governance. Hadoop does not have easy-to-use, full-feature tools for data management, data cleansing, governance and metadata. Especially lacking are tools for data quality and standardization.

3. How Is Hadoop Being Used

Going beyond its original goal of searching millions (or billions) of web pages and returning relevant results, many organizations are looking on Hadoop as their next big data platform. Popular uses today include the following.

3.1 Low-Cost Storage and Data Archive

The modest cost of commodity hardware makes Hadoop useful for storing and combining data such as transactional, social media, sensor, machine, scientific, click streams, etc. The low-cost storage lets you keep information that is not deemed currently critical but that you might want to analyze later.

3.2 Sandbox for Discovery and Analysis

Because Hadoop is designed to deal with volumes of data in a variety of shapes and forms, it can run analytical algorithms. Big data analytics on Hadoop can help your organization operate more efficiently, uncover new opportunities and derive next-level competitive advantage. The sandbox approach provides an opportunity to innovate with minimal investment.

3.3 Data Lake

Data lakes support storing data in its original or exact format. The goal is to offer a raw or unrefined view of data to data scientists and analysts for discovery and analytics. It helps them ask new or difficult questions without constraints. Data lakes are not a replacement for data warehouses.

In fact, how to secure and govern data lakes is a huge topic for IT. They may rely on data federation techniques to create a logical data structure.

3.4 Complement Your Data Warehouse

We're now seeing Hadoop beginning to sit beside data warehouse environments, as well as certain data sets being offloaded from the data warehouse into Hadoop or new types of data going directly to Hadoop. The end goal for every organization is to have a right platform for storing and processing data of different schema, formats, etc., to support different use cases that can be integrated at different levels.

3.5 IoT and Hadoop

Things in the IoT need to know what to communicate and when to act. At the core of the IoT is a streaming, always on torrent of data. Hadoop is often used as the data store for millions or billions of transactions. Massive storage and processing capabilities also allow you to use Hadoop as a sandbox for discovery and definition of patterns to be monitored for prescriptive instruction. You can then continuously improve these instructions, because Hadoop is constantly being updated with new data that doesn't match previously defined patterns.

4. Building a Recommendation Engine in Hadoop

One of the most popular analytical uses by some of Hadoop's large adopters is for web-based recommendation systems. These systems (Facebook, LinkedIn, eBay) analyze huge amounts of data in real time to quickly predict preferences before customers leave the web page.

How: A recommender system can generate a user profile explicitly (by querying the user) and implicitly (by observing the user's behavior) — then compares this profile to reference characteristics (observations from an entire community of users) to provide relevant recommendations. SAS provides a number of techniques and algorithms for creating a recommendation system, ranging from basic distance measures to matrix factorization and collaborative filtering — all of which can be done within Hadoop.

5. Hadoop Core Modules

Currently, four core modules are included in the basic framework from the Apache Foundation:

Hadoop Common — the libraries and utilities used by other Hadoop modules.

Hadoop Distributed File System (HDFS) — the Java-based scalable system that stores data across multiple machines without prior organization.

YARN (Yet Another Resource Negotiator) — provides resource management for the processes running on Hadoop.

MapReduce — a parallel processing software framework. It is comprised of two steps. Map step is a master node that takes inputs and partitions them into smaller sub-problems and then distributes them to worker nodes. After the map step has taken place, the master node takes the answers to all of the sub-problems and combines them to produce output.

6. Commercial Hadoop Distributions

Open-source software is created and maintained by a network of developers from around the world. It's free to download, use and contribute to, though more and more commercial versions of Hadoop are becoming available (these are often called "distros") .With distributions from software vendors, you pay for their version of the Hadoop framework and receive additional capabilities related to security, governance, SQL and management/administration consoles, as well as training, documentation and other services.

7. Getting Data into Hadoop

Here are just a few ways to get your data into Hadoop.

- Use third-party vendor connectors (like SAS/ACCESS® or SAS Data Loader for Hadoop).
- Use Sqoop to import structured data from a relational database to HDFS, Hive and HBase. It can also extract data from Hadoop and export it to relational databases and data warehouses.
- Use Flume to continuously load data from logs into Hadoop.
- Load files to the system using simple Java commands.
- Create a cron job to scan a directory for new files and "put" them in HDFS as they show up. This is useful for things like downloading email at regular intervals.
- Mount HDFS as a file system and copy or write files there.

New Words

commodity	[kə'mɒdɪtɪ]	*n.* 商品；日用品；有价值的物品
enormous	[ɪ'nɔːməs]	*adj.* 巨大的，庞大的
consideration	[kənˌsɪdə'reɪʃn]	*n.* 考虑，关心
protect	[prə'tekt]	*vt.* 保护
failure	['feɪljə]	*n.* 故障，失效
redirect	[ˌriːdi'rekt]	*vt.* 改变方向，改变线路；重新定向
flexibility	[ˌfleksə'bɪlɪtɪ]	*n.* 柔韧性，灵活性；伸缩性
scalability	[skeɪlə'bɪlɪtɪ]	*n.* 可扩展性
independent	[ˌɪndɪ'pendənt]	*adj.* 独立的，不相关连的
iterative	['ɪtərətɪv]	*adj.* 重复的，反复的，迭代的
intercommunicate	[ˌɪntəkə'mjuːnɪkeɪt]	*v.* 互相联络，互相通信
shuffle	['ʃʌfl]	*vt.* 洗牌；搬移

inefficient	[ˌɪnɪ'fɪʃnt]	*adj.* 低效的，无效率的，无能的
productive	[prə'dʌktɪv]	*adj.* 富有成效的；多产的；生产性的；具有创造性的
kernel	['kɜːnl]	*n.* 核；核心；要点 *v.* 把……包在核内
fragmented	[fræg'mentɪd]	*adj.* 成碎片的，片断的
full-fledged	['fʊl'fledʒd]	*adj.* 羽毛生齐的；经过充分训练的，成熟的
scientific	[ˌsaɪən'tɪfɪk]	*adj.* 科学的；有系统的
sandbox	['sændbɒks]	*n.* 沙箱，沙盒
shape	[ʃeɪp]	*n.* 形状；模型；状态
competitive	[kəm'petɪtɪv]	*adj.* 竞争的，比赛的；（价格等）有竞争力的
investment	[ɪn'vestmənt]	*n.* 投资，投资额
unrefined	[ˌʌnrɪ'faɪnd]	*adj.* 未精炼的，未精制的
replacement	[rɪ'pleɪsmənt]	*n.* 代替，替代者
complement	['kɒmplɪment]	*n.* 补充，补充物 *vt.* 补足，补充
offload	[ˌɒf'ləʊd]	*v.* 卸下，卸货
communicate	[kə'mjuːnɪkeɪt]	*vi.* 通讯，通信
torrent	['tɒrənt]	*n.* 爆发，迸发；连续不断
monitor	['mɒnɪtə]	*vt.* 搜集，记录；监控，监听，监视 *n.* 显示器；监测仪
adopter	[ə'dɒptə]	*n.*（新技术）采用者
explicitly	[ɪk'splɪsɪtlɪ]	*adv.* 显式地，明白地，明确地
implicitly	[ɪm'plɪsɪtlɪ]	*adv.* 隐式地；含蓄地；暗示地
matrix	['meɪtrɪks]	*n.* 矩阵
factorization	[ˌfæktərai'zeɪʃən]	*n.* 因数分解
collaborative	[kə'læbərətɪv]	*adj.* 合作的，协作的
module	['mɒdjuːl]	*n.* 模块；组件
parallel	['pærəlel]	*adj.* 平行的，并行的 *adv.* 平行地，并列地
comprise	[kəm'praɪz]	*vt.* 包含，包括；由……组成，由……构成
partition	[pɑː'tɪʃn]	*n.* 划分，分开；分割；隔离物 *vt.* 分开，隔开；区分；分割
sub-problem	[sʌb'prɒbləm]	*n.* 子问题
distribute	[dɪ'strɪbjuːt]	*vt.* 分布，分配，散布，散发，分发
download	[ˌdaʊn'ləʊd]	*v.* 下载
console	[kən'səʊl]	*n.* 控制台，操纵台

connector	[kə'nektə]	*n.* 连接器，连接体
continuously	[kən'tɪnjʊəslɪ]	*adv.* 连续不断地，接连地
command	[kə'mɑːnd]	*n.* 命令，指令 *vt.* 控制，命令；命令 *vi.* 命令，指令 *adj.* 指挥的，根据命令（或要求）而作的
interval	['ɪntəvl]	*n.* 间隔
mount	[maʊnt]	*v.* 挂载；安装，架置

Phrases

open-source software	开源软件
processing power	处理能力
distributed computing model	分布式计算模型
fault tolerance	容错（性）
hardware failure	硬件故障，硬件失效
file intensive	文件密集型
low-level knowledge	底层知识
modest cost	适度成本
click stream	点击流
minimal investment	最低投资
data federation techniques	数据联合技术
recommendation engine	推荐引擎
recommendation system	推荐系统
user profile	用户描述文件
parallel processing	并行处理，多重处理
contribute to ...	为……出贡献

Abbreviations

IoT (Internet of Things)	物联网
IT (Information Technology)	信息技术
HDFS (Hadoop Distributed File System)	Hadoop 分布式文件系统
YARN (Yet Another Resource Negotiator)	另一个资源协调器

Exercises

【Ex.1】Fill in the following blanks according to the text.

1. Hadoop is an ________ for storing data and running applications on clusters of commodity hardware. It provides __________ for any kind of data,__________ and the ability to handle virtually ________ or jobs.
2. Data and application processing are protected against _________. If a node goes down, jobs are automatically redirected to ____________ to make sure ____________.
3. MapReduce is good for __________ and problems that can be divided into ___________, but it's not efficient for __________ and __________.
4. The modest cost of commodity hardware makes Hadoop useful for storing and combining data such as __________, __________, __________, machine, scientific, ________, etc.
5. Because Hadoop is designed to deal with ____________ in a variety of ____________ and ____________, it can run analytical algorithms.
6. Data lakes support storing data ____________ or _____________ format. The goal is to offer _____________ of data to data scientists and analysts for discovery and analytics.
7. Massive storage and processing capabilities also allow you to use Hadoop as ___________ for discovery and ___________ to be monitored for ____________.
8. One of the most popular analytical uses by some of Hadoop's large adopters is for ________. These systems (Facebook, LinkedIn, eBay) analyze huge amounts of data __________ to quickly ____________ before customers leave __________.
9. Currently, there are four core modules included in the basic framework from the Apache Foundation. They are __________, __________, __________ and __________.
10. You can use Sqoop to import structured data from __________ to ____________, and ___________. It can also ____________ and export it to relational databases and __________.

【Ex.2】Translate the following terms or phrases from English into Chinese and vice versa.

1. distributed computing model	1.
2. fault tolerance	2.
3. parallel processing	3.
4. recommendation engine	4.
5. user profile	5.
6. *adj.*合作的，协作的	6.
7. *n.*补充，补充物 *vt.*补足，补充	7.
8. *n.*柔韧性，灵活性；伸缩性	8.
9. *v.*互相联络，互相通信	9.
10. *n.*模块；组件	10.

【Ex.3】Translate the following passages into Chinese.

Fault Tolerance

A fault tolerance is a setup or configuration that prevents a computer or network device from failing in the event of an unexpected problem or error. Making a computer or network fault tolerant requires that the user or company to think how a computer or network device may fail and take steps that help prevent that type of failure. Below are some examples of steps that can be taken.

(1) Power Failure

Have the computer or network device running on aUPS. In the event of a power outage, make sure the UPS can notify an administrator and properly turn off the computer after a few minutes if the power is not restored.

(2) Power Surge

If no UPS is connected to the computer or the UPS does not provide surge protection, connected devices are not protected. We recommend a surge protector to help protect in the event of a power surge.

(3) Data Loss

Run backups daily or at least monthly on the computer if important information is stored on it. Create a mirror of the data on an alternate location.

(4) Device or Computer failure

Have a second device, computer, or computer components available in the event of failure to prevent a long down time.

(5) Unauthorized Access

If connected to a network, set up a firewall.

(6) Frequently Check for Updates

Make sure the operating system and any running programs have the latest updates.

(7) Lock Device or Password to Protect Computer

When not in use lock the computer and store the computer or network device in a secure area.

(8) Overload

Set up an alternate computer or network device that can be used as an alternative access point or can share the load either through a load balancing or round robin setup.

(9) Virus

Make sure the computer has updated virus definitions.

【Ex.4】Fill in the blanks with the words given below.

columns　　effectively　　distributed　　unstructured　　volume
layer　　computing　　process　　load　　accuracy

Differences between Apache Hadoop and RDBMS

Unlike Relational Database Management System (RDBMS), we cannot call Hadoop a database, but it is more of a ____1____ file system that can store and process a huge volume of data sets across a cluster of computers.

Hadoop has two major components: HDFS (Hadoop Distributed File System) and MapReduce. The former one is the storage ____2____ of Hadoop which stores huge amounts of data. MapReduce is primarily a programming model which can effectively process the large data sets by converting them into different blocks of data. These blocks are distributed across the nodes on various machines in the cluster.

However, RDBMS is a structured database approach, in which data gets stored in tables in the forms of rows and ____3____. RDBMS uses SQL or Structured Query Language, which can help update and access the data present in different tables. Traditional RDBMS is not competent to be used in storage of a larger amount of data or simply big data.

1. In Terms of Data Volume

Volume means the quantity of data which could be comfortably stored and effectively processed. Relational databases surely work better when the ____4____ is low, probably gigabytes of data. This has been the case for so long in information technology applications, but when the data size has grown to Terabytes or Petabytes, RDBMS isn't competent to ensure the desired results.

On the other hand, Hadoop is the right approach when the need is to handle a bigger data size. Hadoop can be used to process a huge volume of data ____5____ compared to the traditional relational database management systems.

2. Database Architecture

Considering the database architecture, Hadoop works on the components as:

- HDFS, which is the distributed file system of the Hadoop ecosystem.
- MapReduce, which is a programming model that help process huge data sets.
- Hadoop YARN, which helps in managing the ____6____ resources in multiple clusters.

However, the traditional RDBMS will possess data based on the ACID properties, i.e., Atomicity, Consistency, Isolation, and Durability, which are used to maintain integrity and ____7____ in data transactions.

3. Throughput

It is through the total data volume process over a specific time period that the output could be optimized. Relational database management systems are found to be a failure in terms of achieving a higher throughput if the data ____8____ is high, whereas Apache Hadoop Framework does an appreciable job in this regard. This is one major reason why there is an increasing usage of Hadoop in the modern-day data applications than RDBMS.

4. Data Diversity

The diversity of data refers to various types of data processed. There are structured, unstructured, and semi-structured data available now. Hadoop possesses a significant ability to store and ___9___ data of all the above-mentioned types and prepare it for processing. When it comes to processing big volume of unstructured data, Hadoop is the best-known solution.

However, traditional relational databases could only be used to manage structured or semi-structured data in a limited volume. RDBMS fails in managing ___10___ data. However, it is very difficult to fit in data from various sources to any proper structure. So, we can see that Hadoop is the apt solution in handling data diversity than RDBMS.

Text B
Apache Spark

1. What Is Spark

Apache Spark is an open source big data processing framework built around speed, ease of use, and sophisticated analytics. It was originally developed in 2009 in UC Berkeley's AMPLab, and open sourced in 2010 as an Apache project.

Spark has several advantages compared to other big data and MapReduce technologies like Hadoop and Storm.

First of all, Spark gives us a comprehensive, unified framework to manage big data processing requirements with a variety of data sets that are diverse in nature as well as the source of data.

Spark enables applications in Hadoop clusters to run up to 100 times faster in memory and 10 times faster even when running on disk.

Spark lets you quickly write applications in Java, Scala, or Python. It comes with a built-in set of over 80 high-level operators. And you can use it interactively to query data within the shell.

In addition to Map and Reduce operations, it supports SQL queries, streaming data, machine learning and graph data processing. Developers can use these capabilities stand-alone or combine them to run in a single data pipeline use case.

2. Spark Features

Spark takes MapReduce to the next level with less expensive shuffles in the data processing. With capabilities like in-memory data storage and near real-time processing, the performance can be several times faster than other big data technologies.

Spark also supports lazy evaluation of big data queries, which helps with optimization of the steps in data processing workflows. It provides a higher level API to improve developer productivity and a consistent architect model for big data solutions.

Spark holds intermediate results in memory rather than writing them to disk which is very useful especially when you need to work on the same data set multiple times. It's designed to be an execution engine that works both in-memory and on-disk. Spark operators perform external operations when data does not fit in memory. Spark can be used for processing data sets that larger than the aggregate memory in a cluster.

Spark will attempt to store as much as data in memory and then will spill to disk. It can store part of a data set in memory and the remaining data on the disk. You have to look at your data and use cases to assess the memory requirements. With this in-memory data storage, Spark comes with performance advantage.

Other Spark features include:

- Supports more than just Map and Reduce functions.
- Optimizes arbitrary operator graphs.
- Lazy evaluation of big data queries which helps with the optimization of the overall data processing workflow.
- Provides concise and consistent APIs in Scala, Java and Python.
- Offers interactive shell for Scala and Python. This is not available in Java yet.

Spark is written in Scala Programming Language and runs on Java Virtual Machine (JVM) environment. It currently supports the following languages for developing applications using Spark:

- Scala
- Java
- Python
- Clojure
- R

3. Spark Ecosystem

Other than Spark Core API, there are additional libraries that are part of the Spark ecosystem and provide additional capabilities in Big Data analytics and Machine Learning areas.

These libraries include:

- Spark Streaming: Spark Streaming can be used for processing the real-time streaming data. This is based on micro batch style of computing and processing. It uses the DStream, which is basically a series of RDDs, to process the real-time data.
- Spark SQL: Spark SQL provides the capability to expose the Spark data sets over JDBC API and allows running the SQL like queries on Spark data using traditional BI and visualization tools. Spark SQL allows the users to ETL their data from different formats it's currently in (like JSON, Parquet, a Database), transform it, and expose it for ad-hoc querying.
- Spark MLlib: MLlib is Spark's scalable machine learning library consisting of common learning algorithms and utilities, including classification, regression, clustering, collaborative filtering, dimensionality reduction, as well as underlying optimization primitives.

- Spark GraphX: GraphX is the new (alpha) Spark API for graphs and graph-parallel computation. At a high level, GraphX extends the Spark RDD by introducing the Resilient Distributed Property Graph: A directed multi-graph with properties attached to each vertex and edge. To support graph computation, GraphX exposes a set of fundamental operators (e.g., subgraph, join Vertices, and aggregate Messages) as well as an optimized variant of the Pregel API. In addition, GraphX includes a growing collection of graph algorithms and builders to simplify graph analytics tasks.

4. Spark Architecture

Spark architecture includes the following three main components: Data storage, API, and Resource management.

4.1 Data Storage

Spark uses HDFS file system for data storage purposes. It works with any Hadoop compatible data source including HDFS, HBase, Cassandra, etc.

4.2 API

The API provides the application developers to create Spark based applications using a standard API interface. Spark provides API for Scala, Java, and Python programming languages.

4.3 Resource Management

Spark can be deployed as a stand-alone server or it can be on a distributed computing framework like Mesos or YARN.

5. Resilient Distributed Datasets

Resilient Distributed Dataset or RDD is the core concept in Spark framework. Think about RDD as a table in a database. It can hold any type of data. Spark stores data in RDD on different partitions.

They help with rearranging the computations and optimizing the data processing.

They are also fault tolerance because an RDD knows how to recreate and recompute the data sets.

RDDs are immutable. You can modify an RDD with a transformation but the transformation returns you a new RDD whereas the original RDD remains the same.

RDD supports two types of operations: Transformation and Action.

Transformation: Transformations don't return a single value, they return a new RDD. Nothing gets evaluated when you call a Transformation function, it just takes an RDD and returns a new RDD.

Some of the Transformation functions are map, filter, flatMap, groupByKey, reduceByKey, aggregateByKey, pipe, and coalesce.

Action: Action operation evaluates and returns a new value. When an Action function is called on a RDD object, all the data processing queries are computed at that time and the result value is returned.

6. How to Install Spark

You can install Spark on your machine as a stand-alone framework or use one of Spark Virtual Machine (VM) images available from vendors like Cloudera, HortonWorks, or MapR. Or you can also use Spark installed and configured in the cloud (like Databricks Cloud).

7. Conclusions

In this article, we look at how Apache Spark framework helps with big data processing and analytics with its standard API. We also look at how Spark compares with traditional MapReduce implementation like Apache Hadoop. Spark is based on the same HDFS file storage system as Hadoop, so you can use Spark and MapReduce together if you already have significant investment and infrastructure setup with Hadoop.

You can also combine the Spark processing with Spark SQL, Machine Learning and Spark Streaming.

With several integrations and adapters on Spark, you can combine other technologies with Spark. An example of this is to use Spark, Kafka, and Apache Cassandra together where Kafka can be used for the streaming data coming in, Spark be used to do the computation, and finally Cassandra NoSQL database be used to store the computation result data.

But keep in mind, Spark is a less mature ecosystem and needs further improvements in areas like security and integration with BI tools.

New Words

comprehensive	[ˌkɒmprɪˈhensɪv]	*adj.* 广泛的；综合的
unify	[ˈjuːnɪfaɪ]	*vt.* 统一；使联合；使相同；使一致
diverse	[daɪˈvɜːs]	*adj.* 不同的，多种多样的；变化多的；形形色色的
built-in	[ˈbɪlt ɪn]	*adj.* 嵌入的；内置的；固有的 *n.* 内置
pipeline	[ˈpaɪplaɪn]	*n.* 管道；渠道，传递途径
installment	[ɪnˈstɔːlmənt]	*n.* 部分；（丛书杂志等的）一部，一期
architect	[ˈɑːkɪtekt]	*n.* 建筑师，设计师
intermediate	[ˌɪntəˈmiːdɪət]	*adj.* 中间的，中级的 *n.* 中间物，中间人
arbitrary	[ˈɑːbɪtrərɪ]	*adj.* 随意的，随心所欲的

concise	[kən'saɪs]	*adj.* 简明的，简洁的；简约；精炼
consistent	[kən'sɪstənt]	*adj.* 一致的；连续的
ecosystem	['iːkəʊsɪstəm]	*n.* 生态系统
expose	[ɪk'spəʊz]	*vt.* 揭示
dimensionality	[dɪˌmenʃə'nælɪtɪ]	*n.* 维度；度数
resilient	[rɪ'zɪlɪənt]	*adj.* 有弹性的
vertex	['vɜːteks]	*n.* 顶点
fundamental	[ˌfʌndə'mentl]	*adj.* 基础的，基本的，根本的 *n.* 原理，原则，基本，根本，基础
message	['mesɪdʒ]	*n.* 信息；消息 *v.* 给……发消息
compatible	[kəm'pætəbl]	*adj.* 兼容的，相容的
rearrange	[ˌriːə'reɪndʒ]	*vt.* 重新安排，重新布置
recreate	[ˌriːkrɪ'eɪt]	*v.* 重建，重现
recompute	[ˌriːkəm'pjuːt]	*v.* 再计算，重新算
immutable	[ɪ'mjuːtəbl]	*adj.* 不可改变的
configure	[kən'fɪgə]	*v.* 配置；设定
significant	[sɪg'njfɪkənt]	*adj.* 重要的；显著的；有重大意义的
adapter	[ə'dæptə]	*n.* 适配器；改编者

Phrases

graph data processing	图形数据处理
compares with ...	与……比较
lazy evaluation	延后计算，延迟计算
real-time streaming data	实时流数据
visualization tool	可视化工具
consist of ...	由……组成
learning algorithm	学习算法
dimensionality reduction	降维
Resilient Distributed Property Graph	弹性分布式属性图
directed multi-graph	有向多图
graph computation	图形计算
distributed computing framework	分布式计算框架
file storage system	文件存储系统
keep in mind	记住

Abbreviations

JVM（Java Virtual Machine） Java 虚拟机
RDD（Resilient Distributed Datasets） 弹性分布式数据集
JDBC（Java Data Base Connectivity） Java 数据库连接
JSON（JavaScript Object Notation） JS 对象简谱
VM（Virtual Machine） 虚拟机

Exercises

【Ex.5】Fill in the following blanks according to the text.

1. Apache Spark is an open source big data processing framework built around __________, __________, and __________. It was originally developed in 2009 in __________, and open sourced in 2010 as __________.
2. Spark enables applications in Hadoop clusters to run up to __________ in memory and __________ even when running on disk.
3. Spark holds intermediate results __________ rather than __________ which is very useful especially when you need to work on the same data set multiple times. It's designed to be __________ that works both in-memory and on-disk.
4. Spark is written in __________ and runs on __________ environment. It currently supports the following languages for developing applications using Spark: Scala, __________, __________, __________, __________ and __________.
5. Other than Spark Core API, there are additional libraries that are part of the Spark ecosystem and provide additional capabilities in __________ and __________. These libraries include __________, __________, __________ and __________.
6. There are __________ main components the Spark architecture includes. They are __________, __________ and __________.
7. Resilient Distributed Dataset or RDD is __________ in Spark framework. Think about RDD as __________ in a database. It can hold any type of data. Spark stores data in RDD __________.
8. Transformations don't return __________, they return a new RDD. Some of the Transformation functions are map, __________, __________, groupByKey, __________, __________, pipe, and coalesce.
9. You can install Spark on your machine as __________ or use one of Spark Virtual Machine (VM) images available from __________ like Cloudera, HortonWorks, or MapR. Or you can also use Spark installed and configured __________.

10. But keep in mind, Spark is __________ and needs further improvements in areas like ________ and __________.

参考译文

Apache Hadoop

Hadoop 是一个开源软件框架，用于在商用硬件集群上存储数据和运行应用程序。它为任何类型的数据提供海量存储，具有巨大的处理能力并能处理几乎无限的并发任务或作业。

1. 为什么 Hadoop 很重要

- 能够快速存储和处理大量任何类型的数据。随着数据量和种类的不断增加，特别是来自社交媒体和物联网（IoT）数据的增加，这一点成为要考虑的关键因素。
- 计算能力。Hadoop 的分布式计算模型可以快速处理大数据。使用的计算节点越多，拥有的处理能力就越强。
- 容错。在硬件故障时保护数据和应用进程。如果节点发生故障，作业将自动重新定向到其他节点，以确保分布式计算不会失败。能够自动存储所有数据的多个副本。
- 灵活性。与传统的关系数据库不同，不必在存储数据之前对其进行预处理。可以根据需要存储尽可能多的数据，并决定以后如何使用它。这包括非结构化数据（如文本、图像和视频）。
- 低成本。开源框架是免费的，使用商用硬件存储大量数据。
- 可扩展性。只需添加节点，就可以轻松扩展系统以处理更多数据。只需很少的管理。

2. 使用 Hadoop 有哪些挑战

MapReduce 编程并不适合所有问题。它对于简单的信息请求和可以分成独立单元的问题有用，但对迭代和交互式分析任务效率不高。MapReduce 是文件密集型的。由于除了通过排序和混洗之外，节点不相互通信，因此迭代算法需要多个 map-shuffle / sort-reduce 阶段才能完成。这会在 MapReduce 阶段之间创建多个文件，对于高级分析计算来说效率很低。

有一个广为人知的问题是人才缺乏。很难找到具有足够 Java 技能又能有效使用 MapReduce 的入门级程序员。这就是分销商正在竞相将关系（SQL）技术置于 Hadoop 之上的原因之一。找到具有 SQL 技能的程序员比找到具有 MapReduce 技能的程序员要容易得多。而且，Hadoop 管理似乎是部分艺术和部分科学的，需要掌握操作系统、硬件和 Hadoop 内核设置的基础知识。

数据安全。另一个挑战是围绕分散数据的安全问题，尽管新的工具和技术正在出现。Kerberos 身份验证协议是保证 Hadoop 环境安全的重要一步。

成熟的数据管理和治理。Hadoop 没有针对数据管理、数据清理、治理和元数据的易用的全功能工具，特别缺乏的是数据质量和标准化的工具。

3. 如何使用 Hadoop

最初的目标是搜索数百万（或数十亿）网页并返回相关结果，现在已经超越了。许多组织都希望将 Hadoop 作为他们的下一个大数据平台。如今的流行用途包括以下几方面。

3.1 低成本存储和数据存档

商用硬件的适度成本使 Hadoop 可用于存储和组合交易、社交媒体、传感器、机器、科学、点击流等数据。低成本存储可保留当前不被视为关键但可能稍后想要分析的信息。

3.2 用于发现和分析的沙箱

由于 Hadoop 旨在处理各种各样的大量数据，因此它可以运行分析算法。Hadoop 上的大数据分析可以帮助组织更有效地运营、发现新的机会并获得下一级竞争优势。沙箱方法提供了以最小投资进行创新的机会。

3.3 数据湖

数据湖支持以原始或精确格式存储数据。目标是向数据科学家和分析师提供原始或未提炼的数据视图，以用于科学发现和分析。它可以帮助人们在没有限制的情况下提出新问题。数据湖不是数据仓库的替代品。事实上，如何保护和管理数据湖是 IT 的一个重要话题。人们可能依赖数据联合技术来创建逻辑数据结构。

3.4 补充数据仓库

我们现在看到 Hadoop 开始出现在数据仓库旁边，看到某些数据集从数据仓库卸载而进入 Hadoop，或直接在 Hadoop 使用新类型数据。每个组织的最终目标是拥有一个适当的平台来存储和处理不同模式、格式的数据，以支持可以在不同级别集成的不同用例。

3.5 物联网和 Hadoop

物联网中的物体需要通信并适时动作。物联网的核心是流，总是在数据洪流上。Hadoop 通常用作数百万或数十亿交易的数据存储。大规模存储和处理功能还允许将 Hadoop 用作沙箱，以便发现和定义要监控的规定指令模式。然后，可以不断改进这些指令，因为 Hadoop 会不断更新与先前定义的模式不匹配的新数据。

4. 在 Hadoop 中构建推荐引擎

Hadoop 的大用户最常用的分析用途之一是基于 Web 的推荐系统。这些系统（Facebook、LinkedIn、eBay、Hulu）实时分析大量数据，以便在客户离开网页之前快速预测其偏好。

方法：推荐系统可以显式（通过查询用户）和隐式（通过观察用户的行为）生成用户描述文件——然后将此描述文件与参考特征（来自整个用户社区的观察结果）进行比较，以提供相关建议。SAS 提供了许多用于创建推荐系统的技术和算法，范围从基本距离测量到矩阵分解和协同过滤——所有这些都可以在 Hadoop 中完成。

5. Hadoop 核心模块

目前，Apache Foundation 的基本框架中包含四个核心模块。

Hadoop Common——其他 Hadoop 模块使用的库和实用程序。

Hadoop 分布式文件系统（HDFS）——基于 Java 的可扩展系统，可在不事先组织的情况下跨多台计算机存储数据。

YARN（另一个资源协调器）——为在 Hadoop 上运行的进程提供资源管理。

MapReduce——一个并行处理软件框架。它由两个步骤组成。映射步骤是一个主节点，它接收输入并将它们分为较小的子问题，然后将它们分配给工作节点。在映射步骤完成之后，主节点获取所有子问题的答案并将它们组合以产生输出。

6. 商业 Hadoop 发行版

开源软件由来自世界各地的开发人员在网上创建和维护。它可以免费下载、使用和更新，虽然有越来越多的商业版本可用（这些版本通常称为“发行版”）。使用软件供应商的发行版，需要为其 Hadoop 框架版本付费才能得到与安全、治理、SQL 和管理/管理控制台以及培训、文档和其他服务相关的性能。

7. 将数据导入 Hadoop

以下是将数据导入 Hadoop 的几种方法。

- 使用第三方供应商连接器（如 SAS /ACCESS®或 SAS Data Loader for Hadoop）。
- 使用 Sqoop 将关系数据库中的结构化数据导入 HDFS、Hive 和 HBase。Sqoop 还可以从 Hadoop 中提取数据并将其导出到关系数据库和数据仓库。
- 使用 Flume 将日志中的数据连续加载到 Hadoop 中。
- 使用简单的 Java 命令将文件加载到系统。
- 创建一个 cron 作业来扫描目录中的新文件，并在它们出现时将它们“放入”HDFS 中。这对于定期下载电子邮件等内容非常有用。
- 将 HDFS 挂载为文件系统，并在其中复制或写入文件。

Unit 11

Text A
Big Data Visualization

Big data visualization calls to mind the old saying: "A picture is worth a thousand words." That's because an image can often convey "what's going on" more quickly, more efficiently, and often more effectively than words. Big data visualization techniques exploit this fact: They are all about turning data into pictures by presenting data in pictorial or graphical format. This makes it easy for decision-makers to take in vast amounts of data at a glance to "see" , what it is that the data has to say.

1. What Is Big Data Visualization

Big data visualization involves the presentation of data of almost any type in a graphical format that makes it easy to understand and interpret. But it goes far beyond typical corporate graphs, histograms and pie charts to more complex representations like heat maps and fever charts, enabling decision makers to explore data sets to identify correlations or unexpected patterns.

A defining feature of big data visualization is scale. Today's enterprises collect and store vast amounts of data that would take years for a human to read, let alone understand them. But researchers have determined that the human retina can transmit data to the brain at a rate of about 10 megabits per second. Big data visualization relies on powerful computer systems to ingest raw corporate data and process it to generate graphical representations that allow humans to take in and understand vast amounts of data in seconds.

2. Importance of Big Data Visualization

The amount of data created by corporations around the world is growing every year, and thanks to innovations such as the Internet of Things this growth shows no sign of abating. The problem for businesses is that this data is only useful if valuable insights can be extracted from it and acted upon.

To do that decision makers need to be able to access, evaluate, comprehend and act on data in near real-time, and big data visualization promises a way to be able to do just that. Big data visualization is not the only way for decision makers to analyze data, but big data visualization techniques offer a fast and effective way to:

- Review large amounts of data — Data presented in graphical form enables decision makers to take in large amounts of data and gain an understanding of what it means very quickly, far more quickly than poring over spreadsheets or analyzing numerical tables.
- Spot trends — Time-sequence data often captures trends, but spotting trends hidden in data is notoriously hard, especially when the sources are diverse and the quantity of data is large. But the use of appropriate big data visualization techniques can make it easy to spot these trends, and in business terms a trend that is spotted early is an opportunity that can be acted upon.
- Identify correlations and unexpected relationships — One of the huge strengths of big data visualization is enabling users to explore data sets, not to find answers to specific questions, but to discover what unexpected insights the data can reveal. This can be done by adding or removing data sets, changing scales, removing outliers, and changing visualization types. Identifying previously unsuspected patterns and relationships in data can provide businesses with a huge competitive advantage.
- Present the data to others — An oft-overlooked feature of big data visualization is that it provides a highly effective way to communicate any insights that it surfaces to others. That's because it can convey meaning very quickly and in a way that it is easy to understand: Precisely what is needed in both internal and external business presentations.

3. How Data Visualization Works

The human brain has evolved to take in and understand visual information, and it excels at visual pattern recognition. It is this ability that enables humans to spot signs of danger, as well as to recognize human faces and specific human faces such as family members.

Big data visualization techniques exploit this by presenting data in visual form so it can be processed by this hard-wired human ability almost instantly — rather than, for example, by mathematical analysis that has to be learned and laboriously applied.

The trick with big data visualization is choosing the most effective way to visualize the data to surface any insights it may contain. In some circumstances simple business tools such as pie charts or histograms may reveal the whole story, but with large, numerous and diverse data sets more esoteric visualization techniques may be more appropriate. Various big data visualization examples include:

- Linear: Lists of items, items sorted by a single feature.
- 2D/Planar/geospatial: Cartograms, dot distribution maps, proportional symbol maps, contour maps.
- 3D/Volumetric: 3D computer models, computer simulations.
- Temporal: Timelines, time series charts, connected scatter plots, arc diagrams, circumplex charts.
- Multidimensional: Pie charts, histograms, tag clouds, bar charts, tree maps, heat maps, spider

charts.

• Tree/hierarchical: Dendrograms, radial tree charts, hyperbolic tree charts.

4. Is Big Data Visualization for You

The answer to this question is almost certainly "yes", and here's why. Big data is all about collecting and keeping large amounts of data rather than discarding it. Data storage is cheap but the value of the insights the data contains may be high.

There are a number of ways to analyze data, but the most effective, or indeed the only way, that some insights can be surfaced and exposed is through big data visualization.

In fact, the amount of data that an organization stores does not need to be particularly large in order for it to benefit from big data visualization techniques: The periodic table is a perfect big data visualization example that clearly reveals obscured relationships between just a hundred or so elements.

5. The Challenges to Big Data Visualization

Big data visualization can be an extremely powerful business capability, but before an organization can take advantage of it some key issues need to be addressed. These include:

• Availability of visualization specialists. Many big data visualization tools are designed to be easy enough for anyone in an organization to use, often suggesting appropriate big data visualization examples for the data sets under analysis. But to get the most out of some tools, it may be necessary to employ a specialist in big data visualization techniques who can select the best data sets and visualization styles to ensure the data is exploited to the maximum.

• Visualization hardware resources. Under the hood, big data visualization is essentially a computing task, and the ability to carry out this task quickly — to enable organizations to make decisions in a timely manner using real-time data — may require powerful computer hardware, fast storage systems, or even a move to cloud. That means big data visualization initiatives are as much an IT project as a management project.

• Data quality: The insights that can be drawn from big data visualization are only as accurate as the data that is being visualized. If it is inaccurate or out of date, the value of any insights is questionable. That means people and processes need to be put in place to manage corporate data, metadata, data sources, and any transformations or data cleaning that are performed before storage.

6. Big Data Visualization Tools

A quick survey of the big data tools marketplace reveals the presence of big names including Microsoft, SAP, IBM and SAS. But there are plenty of specialist software vendors offering leading big data visualization tools, and these include Tableau Software, Qlik and TIBCO Software. Leading data visualization products include those offered by:

IBM Cognos Analytics. Driven by their commitment to big data, IBM's analytics package offers a variety of self service options to identify insight more easily.

QlikSense and QlikView. The Qlik solution touts its ability to perform the more complex analysis that finds hidden insights.

Microsoft Power BI. The Power BI tools enable you to connect with hundreds of data sources, then publish reports on the Web and across mobile devices.

Oracle Visual Analyzer. A web-based tool, Visual Analyzer allows creation of curated dashboards to help discover correlations and patterns in data.

SAP Lumira. Calling it "self service data visualization for everyone", Lumira allows you to combine your visualizations into storyboards.

SAS Visual Analytics. The SAS solution promotes its "scalability and governance" along with dynamic visuals and flexible deployment options.

Tableau Desktop. Tableau's interactive dashboards allow you to "uncover hidden insights on the fly", and power users can manage metadata to make the most of disparate data sources.

TIBCO Spotfire. Offers analytics software as a service, and touts itself as a solution that "scales from a small team to the entire organization".

New Words

convey	[kən'veɪ]	*vt.* 传达，传递；表达
exploit	[ɪk'splɔɪt]	*vt.* 采用；利用
pictorial	[pɪk'tɔːrɪəl]	*adj.* 有图片的；图画似的；形象化的 *n.* 画报；画刊；画页
decision-maker	[dɪ'sɪʒn 'meɪkə]	*n.* 决策人，决策者
presentation	[ˌprezn'teɪʃn]	*n.* 提交；演出；陈述，报告
histogram	['hɪstəgræm]	*n.* 柱状图
scale	[skeɪl]	*n.* 规模；比例（尺）；级别 *vt.* 测量
retina	['retɪnə]	*n.* 视网膜
ingest	[ɪn'dʒest]	*vt.* 获取（某事物）；吸收
comprehend	[ˌkɒmprɪ'hend]	*vt.* 理解，领会
spot	[spɒt]	*v.* 认出，发现
notoriously	[nəʊ'tɔːrɪəslɪ]	*adv.* 著名地，众所周知地
appropriate	[ə'prəʊprɪət]	*adj.* 适当的；恰当的；合适的
reveal	[rɪ'viːl]	*vt.* 显露，揭露 *n.* 揭示，展现

unsuspected	[ˌʌnsəˈspektɪd]	*adj.* 未被怀疑的
oft-overlooked	[ˈɒftˌəuvəˈlukt]	*adj.* 视而不见的，忽视的
precisely	[prɪˈsaɪslɪ]	*adv.* 精确地；恰好地
internal	[ɪnˈtɜːnl]	*adj.* 内部的
external	[ɪkˈstɜːnl]	*adj.* 外面的，外部的
excel	[ɪkˈsel]	*vt.* 优于，擅长 *vi* .胜过
hard-wired	[ˈhɒrd ˈwaɪəd]	*adj.* 基本的；无法改变的；硬接线的；硬连线的
laboriously	[ləˈbɔːrɪəslɪ]	*adv.* 费力地；辛勤地；艰苦地
trick	[trɪk]	*n.* 诀窍
surface	[ˈsɜːfɪs]	vi . 浮出水面；显露 vt. 使浮出水面
esoteric	[ˌesəˈterɪk]	adj. 难解的，深奥的；只有内行才懂的
planar	[ˈpleɪnə]	adj. 平面的，平坦的
geospatial	[ˌdʒiːəʊˈspeɪʃl]	*n.* 地理空间
cartogram	[ˈkɑːtəgræm]	*n.* 统计地图
volumetric	[ˌvɒljʊˈmetrɪk]	*adj.* 体积的；容积的；测定体积的
timeline	[ˈtaɪmlaɪn]	*n.* 时间轴，时间表
dendrogram	[ˈdendrəʊgræm]	*n.* 树状图，树形图
discard	[dɪsˈkɑːd]	*vt.* 丢弃，抛弃
cheap	[tʃiːp]	*adj.* 便宜的，廉价的
obscure	[əbˈskjʊə]	*adj.* 不清楚的；隐蔽的 *vt.* 使……模糊不清，掩盖
suggest	[səˈdʒest]	*vt.* 建议，提议；使想起；启示
inaccurate	[ɪnˈækjʊrɪt]	*adj.* 不准确的；不精密的；不正确的；有错误的
questionable	[ˈkwestʃənəbl]	*adj.* 可疑的，有疑问的 *adv.* 可疑地，不真实地
survey	[ˈsɜːveɪ]	*vt.* 调查 *n.* 调查（表）
commitment	[kəˈmɪtmənt]	*n.* 承诺，许诺；委任，委托；致力
tout	[taʊt]	*v.* 兜售；招徕
correlation	[ˌkɒrɪˈleɪʃn]	*n.* 相互关系；相关性
storyboard	[ˈstɔːrɪbɔːd]	*n.*（电影电视节目或商业广告等的）情节串连图板
governance	[ˈgʌvənəns]	*n.* 治理，管理；支配

Phrases

graphical format	图形格式
vast amounts of	巨量
at a glance	看一眼就……，马上
big data visualization	大数据可视化
corporate graph	企业图
pie chart	饼图
heat map	热图；热点图，热力图，热度图
fever chart	热度图
sign of abating	减弱的迹象
numerical table	数字表格
spot trend	发现趋势
time-sequence data	时序数据
dot distribution map	点分布图
proportional symbol map	比例符号图
contour map	等高线图
time series chart	时间序列图
connected scatter plot	连通散点图
arc diagram	圆弧图
circumplex chart	环形图
tag cloud	标签云
bar chart	条形图
tree map	树图
spider chart	蜘蛛图
radial tree chart	径向树形图
hyperbolic tree chart	双曲线树形图
periodic table	周期表
under the hood	在后台，在底层
be drawn from	来自
mobile device	移动设备
along with	和……一起，和……一道；随着
on the fly	即时地，匆忙地；赶紧地

Abbreviations

2D（2-Dimensional） 二维
3D（3-Dimensional） 三维

Exercises

【Ex.1】Answer the following questions according to the text.

1. What does big data visualization call to mind? Why?
2. What does big data visualization rely on powerful computer systems to do?
3. What does data presented in graphical form enable decision makers to do?
4. What is one of the huge strengths of big data visualization?
5. What is an oft-overlooked feature of big data visualization? Why?
6. What is trick with big data visualization?
7. What is the most effective, or indeed the only way, that some insights can be surfaced and exposed?
8. What are many big data visualization tools designed to?
9. What do the Power BI tools enable you?
10. What does the SAS solution promote?

【Ex.2】Translate the following terms or phrases from English into Chinese and vice versa.

1. mobile device	1. ________
2. numerical table	2. ________
3. tag cloud	3. ________
4. time-sequence data	4. ________
5. dendrogram	5. ________
6. *n.*诀窍	6. ________
7. *n.*地理空间	7. ________
8. *n.*视网膜	8. ________
9. *n.*相互关系；相关性	9. ________
10. *vt.*显露，揭露 *n.*揭示，展现	10. ________

【Ex.3】Translate the following passages into Chinese.

Advantages of Big Data Visualization Tools

The primary benefit of big data visualization is that it enables decision makers to better comprehend complex data. Here are some of the most valuable benefits of big data visualization.

1. Quickly Process Large Amounts of Data

It's called big data for a reason. With so much information being captured, organized, and analyzed, there is simply too much raw data for the average human mind to process at any sort of reasonable pace. Big data visualization cuts out the lengthy process of data comprehension, and instead makes it possible for users to digest large, complex data stores at a glance. This is often accomplished through the use of interactive, visually-presented dashboards.

2. Access Important Data in Real Time

In the fast paced world of business, every second counts. Businesses that have to rely on more traditional methods of collating and refining vast amount of data often end up with obsolete information. On the other hand, the best big data visualization tools operate in real time, constantly updating the information being presented, so that at any given time that a user might need to access them, they'll always have the most current data available.

3. Understand Data Better Through Interactivity

Interactive big data visualization tools take the visualization process beyond the use of simple graphs and charts, and allow users the chance to see beyond the numbers. By allowing users the opportunity to delve deeper into causes and trends, big data visualization allows decision makers the chance to see not only what the numbers are saying, but also why.

4. See the Connections Between Business Processes and Performance

Although few leaders would deny that the daily business-processes and operational activities of an organization have a direct impact on overall business-performance, it can be difficult to see exactly how these activities are affecting the company's success. Big data visualization tools solve this problem, by allowing leaders the opportunity to visually discover the hidden relationships that connect the day-to-day processes with business performance.

【Ex.4】Fill in the blanks with the words given below.

efficiently	failure	spreadsheets	messages	interact
complex	sense	visualization	context	reasons

What Is Data Visualization

Data visualization sits atop the big data analytics pyramid and is often the only layer that is visible to executives and other decision-makers. Thus, the success or ____1____ of a big data analytics program often depends on the success or failure of the visualization layer. A company may have the most advanced data capture, storage, and transformation technology and use the most ____2____ algorithms and statistical models to analyze that data, but if the information isn't displayed clearly, accurately and ____3____, the whole point of leveraging big data is lost.

It is often said that data is the new world currency. But let's face it, raw data is boring and difficult to make ____4____ of it in its natural form. Because of the way the human brain processes information, using charts or graphs to visualize large amounts of complex data is easier than poring over ____5____ or reports. According to the Visual Teaching Alliance, studies show that 90% of information transmitted to the brain is visual. Visuals are processed 60,000 times faster in the brain than text, and our eyes can register 36,000 visual ____6____ per hour. It's a no-brainer that visuals work better than text.

All types of organizations are using data ____7____ to help make sense of their data and to comprehend information quickly. Data visualization is a quick, easy way to convey concepts in a universal manner, and due to advances in data visualization technologies, you can experiment with different scenarios by making slight adjustments to available data filters. This is called visual analytics. Visual analytics allows users to directly ____8____ with data, visualize relationships and patterns between operational and marketing activities, gain insight, draw conclusions and make better decisions, quicker. The visibility and clarity delivered by such digital technologies and advanced analytics can give executives unprecedented, granular views into operations. Additionally, it may increase agility and support better strategic decision making by showing the ____9____ why certain recommendations make the most sense. This can have a significant impact on how businesses gain insight from their data.

Data visualization ultimately aims to provide perspective, reveal trends, provide ____10____ and tell a story. Most importantly, data visualization should empower users to harness, in a meaningful way, the power of big data.

Text B
Business Intelligence (BI)

扫码听课文

BI encompasses a wide variety of tools, applications and methodologies that enable organizations to collect data from internal systems and external sources, prepare it for analysis, develop and run queries against that data, and create reports, dashboards and data visualizations to make the analytical results available to corporate decision-makers as well as operational workers.

1. Business Intelligence vs. Data Analytics

Sporadic use of the term business intelligence dates back to at least the 1860s, but consultant Howard Dresner was credited with first proposing it in 1989 as an umbrella term for applying data analysis techniques to support business decision-making processes. What came to be known as BI tools evolved from earlier, often mainframe-based analytical systems, such as decision support systems and executive information systems.

Business intelligence is sometimes used interchangeably with business analytics; in other cases, business analytics is used either more narrowly to refer to advanced data analytics or more broadly to include both BI and advanced analytics.

2. Why Is Business Intelligence Important

The potential benefits of business intelligence tools include accelerating and improving decision-making, optimizing internal business processes, increasing operational efficiency, driving new revenues and gaining competitive advantage over business rivals. BI systems can also help companies identify market trends and spot business problems that need to be addressed.

BI data can include historical information stored in a data warehouse, as well as new data gathered from source systems as it is generated, enabling BI tools to support both strategic and tactical decision-making processes.

Initially, BI tools were primarily used by data analysts and other IT professionals who ran analyses and produced reports with query results for business users. Increasingly, however, business executives and workers are using BI platforms themselves, thanks partly to the development of self-service BI and data discovery tools and dashboards.

3. Types of BI Tools

Business intelligence combines a broad set of data analysis applications, including ad hoc analytics and querying, enterprise reporting, online analytical processing (OLAP), mobile BI, real-time BI, operational BI, cloud and software as a service BI, open source BI, collaborative BI, and location intelligence.

BI technology also includes data visualization software for designing charts and other infographics, as well as tools for building BI dashboards and performance scorecards that display visualized data on business metrics and key performance indicators in an easy-to-grasp way.

Data visualization tools have become the standard of modern BI in recent years. A couple of leading vendors defined the technology early on, but more traditional BI vendors have followed in their path. Now, virtually every major BI tool incorporates features of visual data discovery.

BI programs may also incorporate forms of advanced analytics, such as data mining, predictive analytics, text mining, statistical analysis and big data analytics. In many cases, though, advanced analytics projects are conducted and managed by separate teams of data scientists, statisticians, predictive modelers and other skilled analytics professionals, while BI teams oversee more straightforward querying and analysis of business data.

Business intelligence data is typically stored in a data warehouse or in smaller data marts that hold subsets of a company's information. In addition, Hadoop systems are increasingly being used within BI architectures as repositories or landing pads for BI and analytics data — especially for unstructured data, log files, sensor data and other types of big data.

Before it's used in BI applications, raw data from different source systems must be integrated, consolidated and cleansed using data integration and data quality tools to ensure that users are analyzing accurate and consistent information.

4. BI Trends

In addition to BI managers, business intelligence teams generally include a mix of BI architects, BI developers, business analysts and data management professionals. Business users are also often included to represent the business side and make sure its needs are met in the BI development process.

To help with that, a growing number of organizations are replacing traditional waterfall development with Agile BI and data warehousing approaches. These Agile BI and data warehousing approaches use Agile software development techniques to break up BI projects into small chunks and deliver new functionality to business analysts on an incremental and iterative basis. Doing so can enable companies to put BI features into use more quickly and to refine or modify development plans as business need to change or as new requirements emerge and take priority over earlier ones.

5. BI for Big Data

BI platforms are increasingly being used as front-end interfaces for big data systems. Modern BI software typically offers flexible back-ends, enabling them to connect to a range of data sources. This, along with simple user interfaces, makes the tools a good fit for big data architectures. Users can connect to a range of data sources, including Hadoop systems, NoSQL databases, cloud platforms and more conventional data warehouses, and can develop a unified view of their diverse data.

Because the tools are typically fairly simple, using BI as a big data front end enables a broad number of potential users to get involved rather than the typical approach of highly specialized data architects being the only ones with visibility into data.

6. Advantages of Business Intelligence

Here are top advantages of business intelligence that can help any organization increase profitability by conveying BI system across the firm.

6.1 Authorize Employees

If an organization allows uncomplicated data access for the user which is easy to understand and evaluative, the employees can execute in various ways that can indirectly improve performance and back the entire business plan. Business Intelligence comprises healthy, lively business score registering, investigation, and reporting equipment so that every employee across the firm can make faster and enhanced decisions.

6.2 Unite People to Access Data Competently and Successfully

The initiation of business intelligence has made decision making a lot simpler. Opinion leaders can access and evaluate data at any given point of time and place. The latest information is accessible on the users' desktop or over the internet.

6.3 Simplify Teamwork and Allocation

Business intelligence and partnership expertise enhance managerial efficiency. Firm incorporation of BI enables the employees to share data in a security improved, administered web ambiance with the team members, clients, and associates. They even have a centralized site to supervise their KPIs (key performance indicators), access accounts, evaluate information as well as share texts, and connect to pertinent subject matter.

6.4 Convey Business Intelligence to the Entire Firm

BI backs the width of the firm's business intelligence requirements. Premeditated planning is uncomplicated when familiar equipment is used, data supervision is easier and expansion is more lucrative.

6.5 Examine and Increase Insight

The fully integrated business intelligence tools enable employees to increase insight simply by utilizing well-known and accessible tools. When data is obtained easily and people interact freely, they are better able to investigate and assess information and then make knowledgeable, astute business plans.

6.6 Lessen Training Requirements

With business intelligence people can interconnect with information they desire to access to. Using business tools that are common, easily accessible and extensively backed reduces the training costs of the firm.

6.7 Transport Refined Investigation and Reporting

Impressive scorecard practicality backed by accounts diagrams, graphs and assessments signifies that employees can voluntarily follow KPIs aligned with key business objectives. Accepting and examining the association between KPIs and corporate goals can lead to better comprehension of daily business performance, so that the firm can act on it faster.

7. Disadvantages of Business Intelligence

Some of the major business intelligence disadvantages are as follows.

7.1 Piling of Historical Data

The major objective of business intelligence system is to stockpile past data about a firm's deals and reveal it in such a way that it permits professionals to make decisions. On the flip side, this information generally amounts to a small portion of what the firms actually require to function, besides its restrained worth. While in other situations, the user may not have interest in historical data as many markets that the company regulates are in frequent alteration.

7.2 Cost

Business intelligence at times can be a little too much for small as well as for medium sized enterprises. The use of such system can be expensive for basic business transactions.

7.3 Complexity

Another disadvantage of BI could be its complexity in implementation of data. It can be so intricate that it can make business techniques rigid to deal with. In the view of such premise, many business experts have predicted that these intricacies can ultimately throttle any business.

7.4 Muddling of Commercial Settings

Business intelligence can cause commercial settings to turn out to be much more muddled.

7.5 Limited Use

Like all improved technologies, business intelligence was first established to keep considering the buying competence of affluent firms. Even today BI system cannot be afforded by most of the companies. Although traders in the past few years have started modifying their services towards medium and small sized industries, the fact is that many of such firms do not consider them to be highly essential, for its complexity.

7.6 Time Consuming Implementation

Many firms in today's fast paced industrial scenario are not patient enough to wait for the execution of business intelligence in their organization. It takes around 18 months for data warehousing system to completely implement the system.

Hence, it becomes vital for the firms to give due thought to the business intelligence aspect. Due to the intricacy of these systems, the BI system can create an existence of their own in the firm. It must be understood by the firm that storing data in the business intelligence system just for the sake of it does not increase its worth but results in vice versa effect.

New Words

encompass	[ɪn'kʌmpəs]	*vt.* 包含或包括某事物；围绕，包围
consultant	[kən'sʌltənt]	*n.*（受人咨询的）顾问
mainframe	['meɪnfreɪm]	*n.* 主机
interchangeably	[ɪntə'tʃeɪndʒəblɪ]	*adv.* 可交换地，可交替地
optimize	['ɒptɪmaɪz]	*vt.* 使最优化

rival	[ˈraɪvl]	*n.* 对手；竞争者
		vt. 与……竞争；比得上某人
		vi. 竞争
		adj. 竞争的
infographic	[ɪnfəʊˈgræfɪk]	*n.* 信息图像
scorecard	[ˈskɔːkɑːd]	*n.* 记分卡
conduct	[kənˈdʌkt]	*v.* 组织；实施；执行
mix	[mɪks]	*n.* 混合
		v. 混合；相容
waterfall	[ˈwɔːtəfɔːl]	*n.* 瀑布
chunk	[tʃʌŋk]	*n.* 厚厚的一块；（某物）相当大的数量或部分
priority	[praɪˈɒrɪtɪ]	*n.* 优先，优先权；优先次序
front-end	[frʌnt end]	*n.* 前端
back-end	[bæk end]	*n.* 后端
visibility	[ˌvɪzəˈbɪlɪtɪ]	*n.* 可见性；清晰度
uncomplicated	[ʌnˈkɒmplɪkeɪtɪd]	*adj.* 简单的；不复杂的
teamwork	[ˈtiːmwɜːk]	*n.* 协力；配合；联合作业；集体工作
allocation	[ˌæləˈkeɪʃn]	*n.* 配给，分配
centralized	[sentrəlaɪzd]	*adj.* 集中的，中央的，中心的
pertinent	[ˈpɜːtɪnənt]	*adj.* 有关的；恰当的；关于……的
lucrative	[ˈluːkrətɪv]	*adj.* 获利多的，赚钱的；合算的
voluntarily	[ˈvɒləntərɪlɪ]	*adv.* 志愿地；自动地，自发地
stockpile	[ˈstɒkpaɪl]	*n.* 储备，存储；资源
		vt. 大量贮备
alteration	[ˌɔːlteˈreɪʃn]	*n.* 变化，改变；变更
intricate	[ˈɪntrɪkɪt]	*adj.* 错综复杂的；难理解的
throttle	[ˈθrɒtl]	*v.* 扼杀，压制
muddle	[ˈmʌdl]	*n.* 糊涂，困惑；混乱，杂乱；
		vt. 弄乱，弄糟；使糊涂
		vi. 对付；混日子
affluent	[ˈæfluənt]	*adj.* 富裕的，富足的
		n. 富裕的人
patient	[ˈpeɪʃnt]	*adj.* 有耐性的；能容忍的

Phrases

external source	外部来源，外部资源

business decision-making process	业务决策过程
decision support system	决策支持系统
business analytic	业务分析
executive information system	经理信息系统
operational efficiency	经营效率，运营效率
data discovery tool	数据发现工具
software as a service	软件即服务
location intelligence	智能性定位
easy-to-grasp way	易于掌握的方式
text mining	文本挖掘
statistical analysis	统计分析
opinion leader	意见领袖
share data	共享数据，分享数据
business plan	企业计划，经营计划，业务计划
interconnect with	互连
commercial setting	商业环境

Abbreviation

KPI（key performance indicator）	关键绩效指标

Exercises

【Ex.5】Answer the following questions according to the text.

1. When does sporadic use of the term business intelligence date back to? Who was credited with first proposing it? When?
2. What do the potential benefits of business intelligence tools include?
3. By whom were BI tools primarily used initially?
4. What does business intelligence combine?
5. Where is business intelligence data typically stored?
6. In addition to BI managers, what do business intelligence teams generally include? What are business users also often included to do?
7. What does using BI as a big data front end enable a broad number of potential users to do?
8. If an organization allows uncomplicated data access for the user which is easy to understand and evaluative, what can the employees do?
9. What do the fully integrated business intelligence tools enable employees to do?
10. What are some of the major business intelligence disadvantages?

参考译文

大数据可视化

大数据可视化让人想起一句老话："一张图片胜过千言万语。"这是因为图像通常能够比文字更快、更高效、更有效地传达"正在发生的事情"。大数据可视化技术利用了这一事实：它们都通过图形或图形格式呈现数据。这使得决策者可以一目了然地"看明白"大量数据所表达的事情。

1．什么是大数据可视化

大数据可视化指以图形格式呈现几乎任何类型的数据，使其易于理解和解释。但它远远超出典型的企业图、直方图和饼图，以及更复杂的表示（如热图和热度图），使决策者能够探索数据集以识别相关性或意想不到的模式。

大数据可视化的一个定义性特征是规模。现在的企业收集和存储大量数据，人们读完这些数据尚需数年，更不用说理解了。但研究人员已经确定，人类视网膜能够以大约每秒 10 兆比特的速度向大脑传输数据。大数据可视化依赖于功能强大的计算机系统来获取原始公司数据并对其进行处理，以生成图形，使人们能够在几秒内接收并理解大量数据。

2．大数据可视化的重要性

世界各地的公司创造的数据量每年都在增长，而且由于物联网等创新，这种增长并没有减弱的迹象。企业面临的问题是，只有从中提取有价值的见解并采取相应行动，这些数据才有用。

为此，决策者需要能够近乎实时地访问、评估、理解和处理数据，而大数据可视化承诺能够做到这一点。大数据可视化不是决策者分析数据的唯一方式，但该技术提供了一种快速有效的方法：

- 审查大量数据——以图形形式呈现的数据使决策者能够获取大量数据并快速了解其含义——远比通过电子表格分析或数字表分析更快。
- 发现趋势——时间序列数据通常捕获趋势，但发现隐藏在数据中的趋势非常困难——特别是当数据来源多样且量很大时。但是，使用适当的大数据可视化技术可以很容易地发现这些趋势，并且从业务角度来看，早期发现趋势就有机会采取行动。
- 识别相关性和意想不到的关系——大数据可视化的巨大优势之一是使用户能够探索数据集——不是找到特定问题的答案，而是发现数据可以揭示的意想不到的见解。这可以通过添加或删除数据集、更改比例、删除异常值以及更改可视化类型来完成。识别以前未预料到的数据模式和关系可以为企业提供巨大的竞争优势。
- 向他人展示数据——大数据可视化的一个经常被忽视的特点是它提供了一种高效的方式来向其他人展示任何见解。这是因为它可以非常快速地传达意义，并且易于理解：这正是内部和外部业务演示所需的。

3. 数据可视化如何工作

人类大脑已经发展到能接收和理解视觉信息，并且在视觉模式识别方面表现优异。正是这种能力使人类能够发现危险的迹象，以及识别人类面孔和特定人（如家庭成员）的面孔。

大数据可视化技术通过以可视形式呈现数据来有效利用这一点，因此几乎可以立即通过这种基本的人类能力来处理数据——而不是通过如数学分析来完成，数学分析需要学习而且应用起来也困难。

大数据可视化的诀窍是选择最有效的方式来让数据可视化，以显示其可能包含的任何见解。在某些情况下，简单的业务工具（如饼图或直方图）可能会揭示整个故事，但是对于大量、众多和多样化的数据集，更为深奥的可视化技术可能更合适。各种大数据可视化示例包括：

- 线性：项目列表，按单个要素排序的项目。
- 2D/平面/地理空间：统计地图、点分布图、比例符号图、等高线图。
- 3D/体积：三维计算机模型、计算机模拟。
- 时间：时间轴、时间序列图、连通散点图、圆弧图、环形图。
- 多维：饼图、直方图、标签云、条形图、树图、热图、蜘蛛图。
- 树/层次结构：树形图、径向树形图、双曲线树形图。

4. 大数据可视化可以为你服务吗

这个问题的答案几乎肯定是“是的”。原因是大数据就是收集和保存大量数据（而不是丢弃它）。数据存储很便宜，但数据所包含的见解的价值可能很高。

有许多方法可以分析数据，但最有效——或者实际上唯一的方式——是通过大数据可视化来展示和揭示某些洞察力。

实际上，受益于大数据可视化技术，组织要存储的数据量不需要特别大：元素周期表是一个完美的大数据可视化示例，它清楚地显示了大约一百个元素之间的隐藏关系。

5. 大数据可视化的挑战

大数据可视化可以是一种非常强大的业务功能，但在组织可以利用它之前，需要解决一些关键问题，具体包括：

- 可视化专家。许多大数据可视化工具的设计足以让组织中的任何人使用，通常为正在分析的数据集建议适当的大数据可视化示例。但是为了充分利用某些工具，可能需要聘请大数据可视化技术专家，他们可以选择最佳数据集和可视化样式，以确保最大程度地利用数据。
- 可视化硬件资源。总体而言，大数据可视化本质上是一项计算任务，能够快速执行此任务——使组织能够使用实时数据及时做出决策——可能需要功能强大的计算机硬件、快速存储系统，甚至是迁移到云。这意味着大数据可视化计划既是IT项目，也是管理项目。
- 数据质量。可视化数据的准确性决定了能从大数据可视化中获取的见解：如果数据不准确或过时，则任何见解都值得怀疑。这意味着需要人员和流程来管理企业数据、元数据、数据源以及在存储之前执行的任何转换或数据清理。

6. 大数据可视化工具

快速调查表明大数据工具市场有微软、SAP、IBM 和 SAS 这些知名品牌。但是，有许多专业软件供应商提供领先的大数据可视化工具，其中包括 Tableau Software、Qlik 和 TIBCO Software。领先的数据可视化产品包括以下几种。

IBM Cognos Analytics：致力于对大数据的研究，IBM 的分析软件包提供了多种自助服务选项，可以更轻松地识别有价值的见解。

QlikSense 和 QlikView：Qlik 吹嘘它的解决方案能够执行更复杂的分析，找到隐藏的见解。

Microsoft PowerBI：Power BI 工具能够连接数百个数据源，然后在 Web 上和移动设备上发布报告。

Oracle Visual Analyzer：基于 Web 的工具，Visual Analyzer 允许创建策划仪表板，以帮助发现数据中的相关性和模式。

SAP Lumira：将其称为“每个人的自助服务数据可视化”，Lumira 允许将可视化组合到故事板中。

SAS Visual Analytics：SAS 解决方案推广其“可扩展性和治理”，以及动态可视化和灵活的部署选项。

Tableau Desktop：Tableau 的交互式仪表板允许“即时发现隐藏的见解”，高级用户可以管理元数据以充分利用不同的数据源。

TIBCO Spotfire：提供分析软件即服务，并宣称提供“从小团队扩展到整个组织”的解决方案。

Unit 12

Text A
Data Security

1. Why Is Maintaining Big Data Security So Difficult

More and more data is collected and stored today than ever before, making big data a solution to just about every industry's needs. Customers and clients want solutions and options catered perfectly to their needs before they even know they need it. Silos of data store personal information that allows companies and businesses to personalize interactions and shopping experiences for every individual. But with this great reaping of data comes the difficulty of protecting personal information. Just as companies are becoming smarter and innovating their collection and analysis of big data, hackers are also becoming smarter and innovating their attacks on sensitive and expensive information.

From Target to Home Depot and JPMorgan Chase, big named companies have been hit by hackers, but that doesn't mean smaller companies that also hold your personal information aren't susceptible. In fact, they are sometimes more vulnerable as they don't often have the budget to invest in top notch integrated security solutions. These silos of data that companies store are a goldmine for cyber criminals. Data breaches on companies that collect and store big data are becoming more common and aren't going away anytime soon.

But protecting big data isn't just a matter of throwing up firewalls and using good passwords. Big data comes from a variety of sources such as mobile devices, emails, cloud applications and servers. The more complex and varied the data sets are, the harder it is to protect. A more diverse data collection requires more work to protect it.

For some companies, security spending is still alarmingly low. According to experts, about 10 percent of an IT budget should be spent on security, but at the moment, the average is under 9 percent. Securing big data can be difficult when executives aren't interested or don't understand the importance for providing the necessary funds to invest in big data security.

On top of this, there has been a large gap in the big data skills needed in IT personnel. Many problems in big data security can be solved with limited resources as long as the right people are on it. But many of the open IT security jobs have gone unfilled due to lack of interest and even a lack of applicants, which causes too few data experts on staff, and a higher challenge in addressing security deficiencies. After a lack of available data experts, there is also a large deficit in the knowledge of other employees. Some companies don't regulate access controls within their organization and others are practicing poor computer and information security techniques.

Another big roadblock to data security is the anonymity issue. Many consumers and customers are wary of businesses and companies having access to such personal parts of their lives like behaviors, birth dates, motivations and even who their kids are. Many companies are able to resolve these concerns with policies that mask data and aggregate data sets, although these methods aren't always the most effective. The right equipment handled by the right personnel is needed to put data sets back together in order to re-identify customers.

Similarly to this, there is a huge gap in designed security. Some systems aren't effective on their own, while others can't keep up with the changing pace of data mining tactics. Many big data platforms aren't designed to address security issues. Because of this, most platforms lack encryption, compliance, risk management, policy enabling, and other security features. This requires organizations and companies to have the right teams to build those security features into the platforms themselves.

Unfortunately, there are even more security challenges out there and they are changing each day, making a need for versatility and quick response a must with IT departments. Any companies that work with or handle big data will face these challenges on a daily basis, thus requiring a much-needed push for big data security. But every problem has a solution and knowing just what your weaknesses are is the first step in having better big data security.

2. Challenges to Big Data Security and Privacy

Big data could not be described just in terms of its size. However, to generate a basic understanding, big data are data sets which can't be processed in conventional database ways to their size. This kind of data accumulation helps to improve customer care service in many ways. However, such huge amounts of data can also bring forth many privacy issues, making big data security a prime concern for any organizations. Working in the field of data security and privacy, many organizations are acknowledging these threats and taking measures to prevent them.

2.1 Why Big Data Security Issues Are Surfacing

Big data is nothing new to large organizations. However, it's also becoming popular among smaller and medium sized firms due to cost reduction and provided ease to managing data.

Cloud-based storage has facilitated data mining and collection. However, this big data and cloud storage integration has caused a challenge to privacy and security threats.

The reason for such breaches may also be that security applications that are designed to store certain amounts of data cannot check the big volumes of data that the aforementioned data sets have. Also, these security technologies are inefficient to manage dynamic data and can control static data only. Therefore, just a regular security check can not detect security patches for continuous streaming data. For this purpose, you need full-time privacy during data streaming and big data analysis.

2.2 Protecting Transaction Logs and Data

Data stored in a storage medium, such as transaction logs and other sensitive information, may have varying levels, but that's not enough. For instance, the transfer of data between these levels gives the IT manager insight over the data which is being moved. Data size being continuously increased, the scalability and availability makes auto-tiering necessary for big data storage management. Yet, new challenges are being posed to big data storage as the auto-tiering method doesn't keep track of data storage location.

2.3 Validation and Filtration of End-Point Inputs

End-point devices are the main factors for maintaining big data. Storage, processing and other necessary tasks are performed with the help of input data, which is provided by end-points. Therefore, an organization should make sure to use an authentic and legitimate end-point devices.

2.4 Securing Distributed Framework Calculations and Other Processes

Computational security and other digital assets in a distributed framework mostly lack security protections. The two main prevention for it are securing the mappers and protecting the data in the presence of an unauthorized mapper.

2.5 Securing and Protecting Data in Real Time

Due to large amounts of data generation, most organizations are unable to maintain regular checks. However, it is most beneficial to perform security checks and observation in real time or almost in real time.

2.6 Protecting Access Control Method Communication and Encryption

A secured data storage device is an intelligent step in order to protect the data. Yet, because most often data storage devices are vulnerable, it is necessary to encrypt the access control methods as well.

2.7 Data Provenance

To classify data, it is necessary to be aware of its origin. In order to determine the data origin accurately, authentication, validation and access control could be performed.

2.8 Granular Auditing

Analyzing different kinds of logs could be advantageous and this information could be helpful in recognizing any kind of cyber attack or malicious activity. Therefore, regular auditing can be beneficial.

2.9 Granular Access Control

Granular access control of big data stores by NoSQL databases or the Hadoop Distributed File System requires a strong authentication process and mandatory access control.

2.10 Privacy Protection of Data Stores

Data stores such as NoSQL have many security vulnerabilities, which cause privacy threats. A prominent security flaw is that it is unable to encrypt data during the tagging or logging of data or while distributing it into different groups, when it is streamed or collected.

3. Conclusion

Organizations must ensure that all big data bases are immune to security threats and vulnerabilities. During data collection, all the necessary security protections such as real-time management should be fulfilled. Keeping in mind the huge size of big data, organizations should remember the fact that managing such data could be difficult and requires extraordinary efforts. However, taking all these steps would help maintain consumer privacy.

New Words

security	[sɪ'kjʊərɪtɪ]	*n.* 安全；保护，防护 *adj.* 安全的，保密的
option	['ɒpʃn]	*n.* 选项，选择权
cater	['keɪtə]	*vt.* 满足需要，适合
silo	['saɪləʊ]	*n.* 井；筒仓；地下贮藏库
personalize	['pɜːsənəlaɪz]	*vt.* 个性化，使（某事物）针对个人或带有个人感情
reap	[riːp]	*v.* 收获，获得
sensitive	['sensɪtɪv]	*adj.* 敏感的；易受影响的
susceptible	[sə'septəbl]	*adj.* 易受影响的；易受感染的
prey	[preɪ]	*n.* 受害者，受骗者
goldmine	['gəʊldmaɪn]	*n.* 金矿；金山；财源；宝库
cyber	['saɪbə]	*adj.* 计算机（网络）的，信息技术的

firewall	[ˈfaɪəwɔːl]	*n.* 防火墙 *vt.* 用作防火墙
password	[ˈpɑːswɜːd]	*n.* 口令；密码
email	[ˈiːmeɪl]	*n.* 电子邮件 *vt.* 给……发电子邮件
alarmingly	[əˈlɑːmɪŋlɪ]	*adv.* 让人担忧地
gap	[gæp]	*n.* 缺口；分歧
unfilled	[ˌʌnˈfɪld]	*adj.*（职位）空缺的
deficiency	[dɪˈfɪʃnsɪ]	*n.* 缺乏，不足；缺点，缺陷
poor	[pʊə]	*adj.* 低劣的；贫乏的；匮乏的
roadblock	[ˈrəʊdblɒk]	*n.* 路障；障碍 *vi.* 设置路障
anonymity	[ˌænəˈnɪmɪtɪ]	*n.* 匿名；作者不详；匿名者；无名者
concern	[kənˈsɜːn]	*n.* 顾虑；关心；关系，有关 *vt.* 使关心，使担忧；涉及，关系到
mask	[mɑːsk]	*n.* 掩盖，掩饰 *vi.* 隐瞒，掩饰
aggregate	[ˈægrɪgeɪt]	*vt.* 使聚集，使积聚
handle	[ˈhændl]	*vi.* 处理；操作，操控 *n.* 手柄；句柄
tactic	[ˈtæktɪk]	*n.* 战术，策略，手段
versatility	[ˌvɜːsəˈtɪlɪtɪ]	*n.* 易变；多用途
weakness	[ˈwiːknɪs]	*n.* 弱点，缺点
accumulation	[əˌkjuːmjʊˈleɪʃn]	*n.* 积累；累积量；堆积物
acknowledge	[əkˈnɒlɪdʒ]	*vt.* 承认
threat	[θret]	*n.* 威胁
aforementioned	[əˌfɔːˈmenʃənd]	*adj.* 上述的，前述的
control	[kənˈtrəʊl]	*vt.* 控制；管理
patch	[pætʃ]	*n.* 补丁，补片
log	[lɒg]	*n.* 记录；日志
auto-tiering	[ˈɔːtəʊˈtaɪərɪŋ]	*v.* 自动分级
validation	[ˌvælɪˈdeɪʃn]	*n.* 验证，确认
filtration	[fɪlˈtreɪʃn]	*n.* 过滤；筛选
authentic	[ɔːˈθentɪk]	*adj.* 可信的，可靠的；认证了的
legitimate	[lɪˈdʒɪtɪmət]	*adj.* 合法的，合理的；正规的

prevention	[prɪˈvenʃn]	*n.* 预防；阻止，制止
mapper	[ˈmæpə]	*n.* 映射器；映射程序
intelligent	[ɪnˈtelɪdʒənt]	*adj.* 智能的；聪明的；有智力的
vulnerable	[ˈvʌlnərəbl]	*adj.* 易受攻击的
provenance	[ˈprɒvənəns]	*n.* 来源，起源，出处
authentication	[ɔːˌθentɪˈkeɪʃn]	*n.* 身份验证；认证；证明，鉴定
recognize	[ˈrekəgnaɪz]	*v.* 辨认，识别，承认
malicious	[məˈlɪʃəs]	*adj.* 恶意的，存心不良的；预谋的
activity	[ækˈtɪvɪtɪ]	*n.* 活动
mandatory	[ˈmændətərɪ]	*adj.* 强制的；命令的；受委托的
vulnerability	[ˌvʌlnərəˈbɪlɪtɪ]	*n.* 漏洞，弱点
prominent	[ˈprɒmɪnənt]	*adj.* 突出的
immune	[ɪˈmjuːn]	*adj.* 免疫的；有免疫力的；不受影响的
extraordinary	[ɪkˈstrɔːdɪnərɪ]	*adj.* 非凡的；特别的

Phrases

personalized interaction	个性化互动
shopping experience	购物体验
silo of data	数据井
cyber criminal	网络罪犯
just a matter of	只是……的问题
throw up	匆匆建起
be wary of	留神，谨防，提防
risk management	风险管理，风险管控
in many way	在许多方面
security threat	安全威胁
dynamic data	动态数据
static data	静态数据
storage medium	存储介质
computational security	计算安全性
access control method	访问控制方法
granular access control	粒度访问控制
mandatory access control	强制访问控制
security flaw	安全缺陷
keep in mind	牢记

Exercises

【Ex.1】Fill in the blanks according to the text.

1. Just as companies are becoming smarter and innovating ________ and ________ of big data, hackers are also becoming smarter and innovating ________ on sensitive and expensive information.

2. Big data comes from a variety of sources such as ________, emails, ________ and servers. The more complex and varied the data sets are, ________ it is to protect. A more diverse data collection requires ________ to protect it.

3. But many of the open IT security jobs have gone unfilled due to ________ and even a lack of applicants, which causes ________ on staff, and ________ in addressing security deficiencies.

4. Many consumers and customers are wary of businesses and companies having access to such personal parts of their lives like ________, ________, and even ________.

5. Unfortunately, there are even ________ out there and they are changing each day, making a need for ________ and ________ a must with IT departments.

6. Big data is ________ to large organizations, however, it's also becoming popular among ________ firms due to ________ and provided ease to ________.

7. The reason for such breaches may also be that ________ that are designed to store certain amounts of data cannot check ________ that the aforementioned data sets have.

8. Computational security and other digital assets in a distributed framework mostly lack ________. The two main prevention for it are ________ and ________ in the presence of an unauthorized mapper.

9. To classify data, it is necessary to be aware of ________. In order to determine the data origin accurately, ________, ________ and ________ could be performed.

10. A prominent security flaw is that it is unable to encrypt data during ________ of data or while ________ distributing it into ________, when it is streamed or collected.

【Ex.2】Translate the following terms or phrases from English into Chinese and vice versa.

1.	access control method	1.	
2.	cyber criminal	2.	
3.	security threat	3.	
4.	granular access control	4.	
5.	storage medium	5.	
6.	*n.*身份验证；认证；证明，鉴定	6.	

7. *vt.*使聚集，使积聚 7. ______
8. *adj.*智能的；聪明的；有智力的 8. ______
9. *n.*防火墙 *vt.*用作防火墙 9. ______
10. *n.*口令；密码 10. ______

【Ex.3】Translate the following passages into Chinese.

Cybercrime

Cybercrime is defined as a crime in which a computer is the object of the crime (hacking, phishing, spamming) or is used as a tool to commit an offense. Cybercriminals may use computer technology to access personal information, business trade secrets or use the internet for malicious purposes. Criminals can also use computers for communication and document or data storage. Criminals who perform these illegal activities are often referred to as hackers.

Cybercrime may also be referred to as computer crime.

Common types of cybercrime include online bank information theft, identity theft, online predatory crimes and unauthorized computer access. More serious crimes like cyberterrorism are also of significant concern.

Cybercrime encompasses a wide range of activities, but they can generally be broken into two categories:

(1) Crimes that target computer networks or devices. These types of crimes include viruses and denial-of-service (DoS) attacks.

(2) Crimes that use computer networks to advance other criminal activities. These types of crimes include cyberstalking, phishing and fraud or identity theft.

【Ex.4】Fill in the blanks with the words given below.

security	measures	permission	exchange	critical
divide	threats	monitored	improve	limited

Main Challenges to Big Data Security

Almost all data security issues are caused by the lack of effective measures provided by antivirus software and firewalls. These systems are developed to protect the ___1___ scope of information stored on the hard disk, but Big Data goes beyond hard disks and isolated systems.

1. Nine Big Data Security Challenges

- Most distributed systems' computations have only a single level of protection, which is not recommended.
- Non-relational databases (NoSQL) are actively evolving, making it difficult for ___2___ solutions to keep up with demand.

- Automated data transfer requires additional security ___3___, which are often not available.
- When a system receives a large amount of information, it should be validated to remain trustworthy and accurate; this practice doesn't always occur, however.
- Unethical IT specialists practicing information mining can gather personal data without asking users for ___4___ or notifying them.
- Access control encryption and connections security can become dated and inaccessible to the IT specialists who rely on it.
- Some organizations cannot — or do not — institute access controls to ___5___ the level of confidentiality within the company.
- Recommended detailed audits are not routinely performed on Big Data due to the huge amount of information involved.
- Due to the size of Big Data, its origins are not consistently ___6___ and tracked.

2. How Can Big Data Security Be Improved

Cloud computing experts believe that the most reasonable way to ___7___ the security of Big Data is through the continual expansion of the antivirus industry. A multitude of antivirus vendors who offer a variety of solutions provide a better defense against Big Data security ___8___.

Refreshingly, the antivirus industry is often touted for its openness. Antivirus software providers freely ___9___ information about current Big Data security threats, and industry leaders often work together to cope with new malicious software attacks, providing maximum gains in Big Data security.

Here are some additional recommendations to strengthen Big Data security.

- Focus on application security, rather than device security.
- Isolate devices and servers containing ___10___ data.
- Introduce real-time security information and event management.
- Provide reactive and proactive protection.

Text B
What's the Key to Success for Big Data and AI

扫码听课文

Big data and AI can help the C-suite tremendously, as long as their current limitations are understood. If they are implemented properly, our ability to innovate is limited only by our capacity to imagine how these technologies can benefit us.

For the past few years we've seen big data and machine learning take more of a foothold in companies, with many believing the era of artificial intelligence (AI) is here.

What is clear is that with advances being made on an almost daily basis now, businesses need to prepare for a vastly different future. But, in order to take advantage, some companies may need to make a dramatic adjustment in how they work.

At the moment, for many companies, AI and big data are viewed in a way that limits the potential they have to offer. They are often seen as something that can help cut operational costs, rather than as a fundamental approach for generating increases in productivity, output and improved certainty over corporate direction.

In order for AI and big data to be successful, companies must combine them with business expertise and insight — making it something the C-suite can't ignore.

Expanding the use of these technologies can help business leaders answer some big business and organisational questions, and deep domain expertise is a vital part of that. To ensure businesses get the most out of big data and AI, here are some tips for the C-suite to follow.

1. Stay Ahead of the Curve

The success of today's AI is predicated on big volumes of data. The more data it receives, the smarter it becomes. And businesses continue to generate data at exponential rates.

This combination dramatically improves the ability not only to analyse, asses and predict the traditional areas scrutinised by company leaders — such as revenue, spend and sales — but also to bring together a range of data sets from internal and external sources to generate new sources of insight and prediction. As a result, the C-suite will be able to anticipate potential issues or opportunities that may arise.

Allowing the technology to take care of these steps — which have been the responsibility of number-crunching teams until recently — should enable better and faster decision making. Or even validate decisions the technology suggests leading to an increase in strategic thinking.

2. Review Your Processes

As companies begin to make this shift, one of the initial steps for businesses will be to take an overall review of their organisations to determine which functions would benefit from big data and AI solutions.

There will be business gaps that can be filled. Staff will need to be retrained, and processes will have to be adjusted to take advantage of the new environment that has been created.

One example of how this concept can be put into action is security. Many companies are struggling to deal with the multitude of threats out there at the moment and struggle to establish clear protocols to protect their assets.

MIT has developed an AI based system which detects over 85% of attacks by reviewing data from billions of log files entries each day. Over time it learns attack patterns, continuously improving its ability to predict threats.

Using such systems improves threat detection whilst freeing resources to fill other security-strengthening activities — or any other part of the business that requires them.

3. Flexibility Is Key

The pace of technological changes is staggering and will only continue to gather pace, creating new science, new techniques, new companies and new products.

A major challenge is the ability to identify and then incorporate the best solution for a business, and at the right time to maximise advantage. Now here is this more the case than in the AI and big data domain, where hundreds of start-ups are competing to be the next industry leaders.

Businesses must ensure they have a well-structured architecture framework that enables CIOs to respond with the flexibility required to incorporate the new and replace the old. This way, should something be found not to be working or a better solution is discovered, the leaders can decide to remove or replace it with something that might be a better fit.

4. AI Awareness

Automated approaches and technologies can be applied to most processes, so businesses should look for places they can aid and remove any unnecessary work. AI is increasing its contextual awareness and can sift through large amounts of data, helping to support roles that until now need human teams to execute.

However, AI and big data provide support far in excess of what we can do, because available computing power provides a massive delivery capacity.

There are a number of areas where AI can offer added value, such as contract management, compliance, finance, human resources and supply chain. Through AI, businesses can effectively augment the work of teams by rapidly providing valuable insight. Some standard robotics mechanisms can emulate human interaction with a system, thereby using robotic techniques to deliver in the same way a person can — but quicker and with less mistakes.

Invoicing, financial close, payroll, claims management, order management are several areas where this has proved to be the case. And robotic systems are rapidly increasing their cognitive capabilities.

Rather than being programmed with the rules needed to mimic human activities, the software is advancing through machine learning which means that over time it is acquiring the knowledge required to execute business processes.

We have just begun to scratch the surface of what big data and AI can do for us. Their potential is virtually unlimited. They can create a business foundation with greater efficiency and a better-managed cost base, and even help us see into the future.

However, businesses must remember that these tools are not a replacement for humans. Human nature and intuition remain the unique traits that drive organisations forward in a way that automation simply can't.

New Words

tremendously	[triˈmendəslɪ]	*adv.* 极大地，极端地；极其，非常
limitation	[ˌlɪmɪˈteɪʃn]	*n.* 限制，局限
imagine	[ɪˈmædʒɪn]	*vt.* 设想；想象；料想，猜想 *vi.* 想象；猜想，推测
foothold	[ˈfʊthəʊld]	*n.* 立足处；稳固地位
vastly	[ˈvɑːstlɪ]	*adv.* 极大地；广大地；深远地
adjustment	[əˈdʒʌstmənt]	*n.* 调整
ignore	[ɪgˈnɔː]	*vt.* 忽视，不顾
organisational	[ˌɔːgənaɪˈzeɪʃənl]	*adj.* 组织的
exponential	[ˌekspəˈnenʃl]	*n.* 指数 *adj.* 指数的，幂数的
scrutinise	[ˈskruːtɪnaɪz]	*vt.* 仔细检查 *n.* 仔细或彻底检查
anticipate	[ænˈtɪsɪpeɪt]	*vt.* 预感，预见，预料
responsibility	[rɪˌspɒnsəˈbɪlɪtɪ]	*n.* 责任；职责；责任感，责任心；负责任
retrain	[riːˈtreɪn]	*v.* 再训练；重新教育，再教育
adjust	[əˈdʒʌst]	*v.*（改变……以）适应；调整，校正
multitude	[ˈmʌltɪtjuːd]	*n.* 大量，许多
maximise	[ˈmæksɪmaɪz]	*vt.* 使（某事物）增至最大限度；最大限度地利用（某事物）
start-up	[stɑːt-ʌp]	*n.* 刚成立的公司，新企业；启动
replace	[rɪˈpleɪs]	*vt.* 替换，代替
awareness	[əˈweənɪs]	*n.* 察觉，觉悟，意识
unnecessary	[ʌnˈnesəsərɪ]	*adj.* 不必要的，多余的；无用的
contextual	[kənˈtekstʃʊəl]	*adj.* 前后关系的；与上下文有关的；与语境相关的
sift	[sɪft]	*vi.* 筛；细查
augment	[ɔːgˈment]	*vt.* 增强，加强；增加；（使）扩张，扩大 *n.* 增加，补充物
robotic	[rəʊˈbɒtɪk]	*adj.* 机器人的；自动的
emulate	[ˈemjʊleɪt]	*vt.* 仿真；竞争；努力赶上
cognitive	[ˈkɒgnɪtɪv]	*adj.* 认知的；认识的
unlimited	[ʌnˈlɪmɪtɪd]	*adj.* 无限的；无数的
intuition	[ˌɪntjʊˈɪʃn]	*n.* 直觉

Phrases

C-suite	C 型雇员，最高管理层；企业高管；高管；公司管理层；企业决策层；指企业最高管理层，因其英文名称开头字母都是 C 而得名。
operational cost	经营成本，运营成本
vital part	要害
take care of	处理；应对；照顾
number-crunching team	数字处理团队
strategic thinking	战略思维
major challenge	重大挑战
architecture framework	体系框架，结构框架
human resource	人力资源
supply chain	供应链
claims management	索赔管理
order management	订单管理
unique trait	独有特质

Exercises

【Ex.5】Fill in the blanks according to the text.

1. Big data and AI can help ________ tremendously, as long as their current limitations are understood. For the past few years we've seen ____________ and ____________ take more of a foothold in companies, with many believing the era of __________ is here.

2. At the moment, for many companies, AI and big data are viewed in a way that __________ they have to offer. They are often seen as something that can help __________, rather than as __________ for generating increases in ____________, ____________ and improved certainty over corporate direction.

3. Expanding the use of these technologies can help ____________ answer some big business and organisational questions, and ____________ is a vital part of that.

4. The success of today's AI is predicated on __________. The more data it receives, __________ it becomes. And businesses continue to generate data ____________.

5. Many companies are struggling to deal with ________ out there at the moment and struggle to ________ to protect ________.

6. A major challenge is the ability to ________ and then ________ for a business, and ________ to maximise advantage.

7. Businesses must ensure they have ________ that enables CIOs to respond with ________ required to incorporate the new and ________.

8. There are a number of areas where AI can offer added value, such as ________, compliance, ________, ________ and ________. Through AI, businesses can effectively augment the work of teams by ________.

9. Rather than being programmed with ________ to mimic ________, the software is advancing through machine learning which means that over time it is acquiring ________ required to execute ________.

10. However, businesses must remember that these tools are ________ for humans. Human nature and ________ remain the unique traits that drive organisations forward in a way that ________.

参考译文

数据安全

1. 为什么确保大数据安全如此困难

今天收集和存储的数据比以往任何时候都要多，数据几乎可以解决所有行业的需求问题。顾客和客户需要完全满足其需求的解决方案和选项，甚至希望在他们意识到这种需求之前就已得到满足。数据仓库存储个人信息，允许公司和企业为每人提供个性化交互和购物体验。但是，因为获得的数据量巨大，所以保护个人信息的难度很大。正如公司在大数据收集和分析方面更加智能和不断创新一样，黑客也变得更加聪明，并且他们也不断创新那些攻击敏感且昂贵信息的方法。

从塔吉特公司（Target）到家得宝公司（Home Depot）、摩根大通公司（JPMorgan Chase），这些大名鼎鼎的公司受到了黑客的攻击，但这并不意味着那些持有您个人信息的小公司不容易受影响。实际上，它们有时更易成为受害者，因为它们通常没有预算来投资一流的集成安全解决方案。公司存储的这些数据并是网络犯罪分子的金矿。收集和存储大数据的公司的数据泄露正变得越来越普遍，并且不会很快消失。

但保护大数据不仅仅是设立防火墙和使用好的密码。大数据来源多样，如移动设备、电子邮件、云应用程序和服务器。数据集越复杂多样，保护起来就越困难。更多样化的数据收集需要更多的工作来保护它。

对于一些公司来说，安全支出仍然低得惊人。据专家介绍，大约 10%的 IT 预算应该花在安全上，但目前平均水平低于 9%。当高管不感兴趣或不明白投资大数据安全的重要性时，保护大数据可能很困难。

除此之外，IT 人员所需的大数据技能存在很大差距。只要有合适的人员，大数据安全中的许多问题都可以用有限的资源来解决。但是由于缺乏兴趣甚至缺乏申请人，许多开放的 IT 安全工作空缺，这导致数据专家员工太少，从而在解决安全缺陷方面面临的挑战更大。除了缺乏可用的数据专家，其他员工的知识也存在很大的不足。有些公司没有对其组织内的访问控制进行管理，其他公司则采用不良的计算机和信息安全技术。

数据安全的另一个重大障碍是匿名问题。许多消费者和顾客都担心企业和公司可以访问他们生活中的这些个人区域，如行为、出生日期、动机，甚至他们的孩子。许多公司能够通过掩盖数据和聚合数据集的策略来解决这些问题，尽管这些方法并不总是最有效的。需要由合适的人员用适当的设备将数据集重新组合在一起，以便重新识别客户。

与此类似，在设计的安全性方面存在巨大差距。有些系统本身并不十分有效，而另外一些系统则无法跟上数据挖掘策略不断变化的步伐。许多大数据平台的设计并不是为了解决安全问题。因此，大多数平台缺乏加密、遵从性、风险管理、策略启用和其他安全功能。这就要求组织和公司拥有正确的团队，将这些安全特性构建到平台之中。

不幸的是，外面还有更多的安全挑战，而且它们每天都在发生变化，这使得对 IT 部门的多功能性和快速反应的需求成为必须。任何一家处理或管理大数据的公司每天都将面临这些挑战，这需要大力推动。没有解决不了的问题，知晓您的劣势所在正是确保大数据安全的第一步。

2. 大数据安全和隐私的挑战

大数据无法仅根据其规模来描述。但是，最基本的理解是，大数据是无法以传统数据库方式处理的数据集。这种数据积累有助于以多种方式改善客户服务。但是，如此庞大的数据也会带来许多隐私问题，使大数据安全成为任何组织的主要关注点。在数据安全和隐私领域，许多组织开始认识到这些威胁的存在，并采取措施避免这些威胁。

2.1 为什么大数据安全问题正在浮出水面

对于大型组织而言，大数据并不是什么新鲜事，但由于成本降低和管理数据的便利性，大数据也在中小型企业中流行起来。

基于云的存储促进了数据挖掘和收集。但是，这种大数据和云存储集成对隐私和安全带来了挑战。

造成此类破坏的原因很可能是设计用于存储某些数据量的安全应用程序不能检测上述数据集所具有的大量数据。此外，这些安全技术在管理动态数据方面效率低下，并且只能控制静态数据。因此，仅定期安全检查无法检测连续流数据的安全补丁。为此，需要在数据流和大数据分析时全时段保护隐私。

2.2 保护事务日志和数据

存储在存储介质中的数据（例如事务日志和其他敏感信息）可能具有不同的层级，但这还不够。例如，这些不同层级之间的数据传输使 IT 管理者能够洞察正在移动的数据。随着数据规模的不断增加，其可扩展性和可用性使得必须进行自动分级才能进行大数据存储管理。然而，由于自动分级方法无法跟踪数据存储位置，因此大数据存储面临新的挑战。

2.3 端点输入的验证和过滤

端点设备是维护大数据的主要因素。存储、处理和其他必要任务的执行都离不开输入数据，输入数据由端点提供。因此，组织应确保使用可靠合法的端点设备。

2.4 确保分布式框架计算和其他过程的安全

计算安全性和分布式框架中的其他数字资产大多缺乏安全保护。实时数据安全的两个主要预防措施是对未经授权的映射器中的数据执行映射和保护。

2.5 实时保护和保护数据

由于生成的数据量很大，大多数组织无法维持定期检查。但是，实时或几乎实时地执行安全检查和观察最为有益。

2.6 保护访问控制方法通信和加密

要保护数据，使用安全存储设备是一个明智的办法。然而，由于大多数情况下数据存储设备易受攻击，因此也必须加密访问控制方法。

2.7 数据来源

要对数据进行分类，必须了解其来源。为了准确地确定数据来源，可以使用认证、验证和访问控制。

2.8 粒度审计

分析不同类型的日志可能是有利的，这些信息可能有助于识别任何类型的网络攻击或恶意活动。因此，定期审核是有益的。

2.9 粒度访问控制

用NoSQL数据库或Hadoop分布式文件系统对大数据存储的粒度访问实施控制需要强大的身份验证过程和强制性的访问控制。

2.10 数据存储的隐私保护

NoSQL 等数据存储存在许多安全漏洞，这些漏洞会导致隐私威胁。一个突出的安全漏洞是，在标记或记录数据期间或在流式传输或收集数据时，无法加密数据；把数据分发到不同组的时候，也无法加密数据。

3．结论

组织必须确保所有大数据库都免受安全威胁和漏洞的影响。在数据收集期间，应实现所有必要的安全保护，例如实时管理。考虑到大数据的庞大规模，组织应该记住管理此类数据可能很困难并需要辛苦的付出。但是，采取所有这些步骤将有助于维护消费者隐私。